CAMBRIDGE LIBRARY COLLECTION

Books of enduring scholarly value

Mathematical Sciences

From its pre-historic roots in simple counting to the algorithms powering modern desktop computers, from the genius of Archimedes to the genius of Einstein, advances in mathematical understanding and numerical techniques have been directly responsible for creating the modern world as we know it. This series will provide a library of the most influential publications and writers on mathematics in its broadest sense. As such, it will show not only the deep roots from which modern science and technology have grown, but also the astonishing breadth of application of mathematical techniques in the humanities and social sciences, and in everyday life.

A Treatise on the Calculus of Finite Differences

Self-taught mathematician and father of Boolean algebra, George Boole (1815-1864) published A Treatise on the Calculus of Finite Differences in 1860 as a sequel to his Treatise on Differential Equations (1859). Both books became instant classics that were used as textbooks for many years and eventually became the basis for our contemporary digital computer systems. The book discusses direct theories of finite differences and integration, linear equations, variations of a constant, and equations of partial and mixed differences. Boole also includes exercises for daring students to ponder, and also supplies answers. Long a proponent of positioning logic firmly in the camp of mathematics rather than philosophy, Boole was instrumental in developing a notational system that allowed logical statements to be symbolically represented by algebraic equations. One of history's most insightful mathematicians, Boole is compelling reading for today's student of logic and Boolean thinking.

Cambridge University Press has long been a pioneer in the reissuing of out-of-print titles from its own backlist, producing digital reprints of books that are still sought after by scholars and students but could not be reprinted economically using traditional technology. The Cambridge Library Collection extends this activity to a wider range of books which are still of importance to researchers and professionals, either for the source material they contain, or as landmarks in the history of their academic discipline.

Drawing from the world-renowned collections in the Cambridge University Library, and guided by the advice of experts in each subject area, Cambridge University Press is using state-of-the-art scanning machines in its own Printing House to capture the content of each book selected for inclusion. The files are processed to give a consistently clear, crisp image, and the books finished to the high quality standard for which the Press is recognised around the world. The latest print-on-demand technology ensures that the books will remain available indefinitely, and that orders for single or multiple copies can quickly be supplied.

The Cambridge Library Collection will bring back to life books of enduring scholarly value (including out-of-copyright works originally issued by other publishers) across a wide range of disciplines in the humanities and social sciences and in science and technology.

A Treatise on the Calculus of Finite Differences

GEORGE BOOLE

CAMBRIDGE
UNIVERSITY PRESS

CAMBRIDGE UNIVERSITY PRESS

Cambridge, New York, Melbourne, Madrid, Cape Town, Singapore,
São Paolo, Delhi, Dubai, Tokyo

Published in the United States of America by Cambridge University Press, New York

www.cambridge.org
Information on this title: www.cambridge.org/9781108000925

© in this compilation Cambridge University Press 2009

This edition first published 1860
This digitally printed version 2009

ISBN 978-1-108-00092-5 Paperback

A TREATISE ON THE CALCULUS OF

FINITE DIFFERENCES.

A TREATISE

ON THE

CALCULUS OF FINITE DIFFERENCES.

BY

GEORGE BOOLE, D.C.L.

HONORARY MEMBER OF THE CAMBRIDGE PHILOSOPHICAL SOCIETY;
PROFESSOR OF MATHEMATICS IN THE QUEEN'S UNIVERSITY, IRELAND.

𝕮𝖆𝖒𝖇𝖗𝖎𝖉𝖌𝖊:

MACMILLAN AND CO.

AND 23, HENRIETTA STREET, COVENT GARDEN,

𝕷𝖔𝖓𝖉𝖔𝖓.

1860.

PREFACE.

In the following exposition of the Calculus of Finite Differences, particular attention has been paid to the connexion of its methods with those of the Differential Calculus—a connexion which in some instances involves far more than a merely formal analogy.

Indeed the work is in some measure designed as a sequel to my *Treatise on Differential Equations.* And it has been composed on the same plan.

Mr Stirling, of Trinity College, Cambridge, has rendered me much valuable assistance in the revision of the proof-sheets. In offering him my best thanks for his kind aid, I am led to express a hope that the work will be found to be free from important errors.

GEORGE BOOLE.

Queen's College, Cork,
April 18, 1860.

CONTENTS.

CONTENTS.

FINITE DIFFERENCES.

CHAPTER I.

NATURE OF THE CALCULUS OF FINITE DIFFERENCES.

1. THE Calculus of Finite Differences may be strictly defined as the science which is occupied about the ratios of the simultaneous increments of quantities mutually dependent. The Differential Calculus is occupied about the *limits* to which such ratios approach as the increments are indefinitely diminished.

In the latter branch of analysis if we represent the independent variable by x, any dependent variable considered as a function of x is represented primarily indeed by $\phi(x)$, but, when the rules of differentiation founded on its functional character are established, by a single letter, as u. In the notation of the Calculus of Finite Differences these modes of expression seem to be in some measure blended. The dependent function of x is represented by u_x, the suffix taking the place of the symbol which in the former mode of notation is enclosed in brackets. Thus, if $u_x = \phi(x)$ then

$$u_{x+h} = \phi(x + h),$$
$$u_{\sin x} = \phi(\sin x),$$

and so on. But this mode of expression rests only on a convention, and as it was adopted for convenience, so when convenience demands it is laid aside.

The step of transition from a function of x to its increment, and still further to the *ratio* which that increment bears to the increment of x, may be contemplated apart from its subject, and it is often important that it should be so contemplated, as an operation governed by laws. Let then Δ prefixed to the expression of any function of x, denote the operation of taking the increment of that function corresponding to a given constant increment Δx of the variable x. Then, representing as above the proposed function of x by u_x, we have

$$\Delta u_x = u_{x+\Delta x} - u_x,$$

and
$$\frac{\Delta u_x}{\Delta x} = \frac{u_{x+\Delta x} - u_x}{\Delta x}.$$

Here then we might say that as $\frac{d}{dx}$ is the fundamental operation of the Differential Calculus, so $\frac{\Delta}{\Delta x}$ is the fundamental operation of the Calculus of Finite Differences.

But there is a difference between the two cases which ought to be noted. In the Differential Calculus $\frac{du}{dx}$ is not a true fraction, nor have du and dx any distinct meaning as symbols of quantity. The fractional form is adopted to express the limit to which a true fraction approaches. Hence $\frac{d}{dx}$, and not d, there represents a real operation. But in the Calculus of Finite Differences $\frac{\Delta u_x}{\Delta x}$ is a true fraction. Its numerator Δu_x stands for an actual magnitude. Hence Δ might itself be taken as the fundamental operation of this Calculus, always supposing the actual value of Δx to be given; and the Calculus of Finite Differences might, in its symbolical character, be defined either as the science of the laws of the operation Δ, the value of Δx being supposed given, or as the science of the laws of the operation $\frac{\Delta}{\Delta x}$. In consequence of the fundamental difference above noted between the Differential Calculus and the Calculus of Finite Differences, the term Finite ceases to be necessary as a mark of distinction. The former is a calculus of *limits*, not of *differences*.

2. Though Δx admits of any constant value, the value usually given to it is unity. There are two reasons for this.

First, the Calculus of Finite Differences has for its chief subject of application the terms of series. Now the law of a

series, however expressed, has for its ultimate object the determination of the values of the successive terms as dependent
upon their numerical order and position. Explicitly or implicitly, each term is a *function* of the integer which expresses its position in the series. And thus, to revert to
language familiar in the Differential Calculus, the independent variable admits only of integral values whose common difference is unity. In the series of terms

$$1^2, \ 2^2, \ 3^2, \ 4^2, \ldots$$

the general or x^{th} term is x^2. It is an explicit function of x,
but the values of x are the series of natural numbers, and
$\Delta x = 1$.

Secondly. When the general term of a series is a function
of an independent variable t whose successive differences are
constant but not equal to unity, it is always possible to
replace that independent variable by another, x, whose common difference shall be unity. Let $\phi(t)$ be the general term
of the series, and let $\Delta t = h$; then assuming $t = hx$ we have
$\Delta t = h\Delta x$, whence $\Delta x = 1$.

Thus it suffices to establish the rules of the calculus on the
assumption that the finite difference of the independent
variable is unity. At the same time it will be noted that this
assumption reduces to equivalence the symbols $\dfrac{\Delta}{\Delta x}$ and Δ.

We shall therefore in the following chapters develope the
theory of the operation denoted by Δ and defined by the
equation

$$\Delta u_x = u_{x+1} - u_x.$$

But we shall where convenience suggests consider the more
general operation

$$\frac{\Delta u_x}{\Delta x} = \frac{u_{x+h} - u_x}{h},$$

where $\Delta x = h$.

CHAPTER II.

DIRECT THEOREMS OF FINITE DIFFERENCES.

1. THE operation denoted by Δ is capable of repetition. For the difference of a function of x, being itself a function of x, is subject to operations of the same kind.

In accordance with the algebraic notation of indices, the difference of the difference of a function of x, usually called the second difference, is expressed by attaching the index 2 to the symbol Δ. Thus

$$\Delta\Delta u_x = \Delta^2 u_x.$$

In like manner

$$\Delta\Delta^2 u_x = \Delta^3 u_x,$$

and generally

$$\Delta\Delta^{n-1} u_x = \Delta^n u_x \dots\dots\dots (1),$$

the last member being termed the nth difference of the function u_x. If we suppose $u_x = x^3$, the successive values of u_x with their successive differences of the first, second, and third orders will be represented in the following scheme:

Values of x	1	2	3	4	5	6 ...
u_x	1	8	27	64	125	216 ...
Δu_x	7	19	37	61	91 ...	
$\Delta^2 u_x$	12	18	24	30 ...		
$\Delta^3 u_x$	6	6	6 ...			

It may be observed that each sum of differences may either be formed from the preceding sum by successive subtractions in accordance with the definition of the symbol Δ, or calculated from the general expressions for Δu, $\Delta^2 u$, &c. by assigning to x the successive values 1, 2, 3, &c. Since $u_x = x^3$, we shall have

$$\Delta u_x = (x+1)^3 - x^3 = 3x^2 + 3x + 1,$$

$$\Delta^2 u_x = \Delta (3x^2 + 3x + 1) = 6x + 6,$$

$$\Delta^3 u_x = 6.$$

It may also be noted that the third differences are here constant. And generally *if u_x be a rational and integral function of x of the nth degree, its nth differences will be constant.* For let

$$u_x = ax^n + bx^{n-1} + \&c.,$$

then

$$\Delta u_x = a(x+1)^n + b(x+1)^{n-1}$$
$$- ax^n - bx^{n-1}$$
$$= anx^{n-1} + b_1 x^{n-2} + b_2 x^{n-3} + \&c.,$$

b_1, b_2, &c., being constant coefficients. Hence Δu_x is a rational and integral function of x of the degree $n-1$. Repeating the process, we have

$$\Delta^2 u_x = an(n-1)x^{n-2} + C_1 x^{n-3} + C_2 x^{n-4} + \&c.,$$

a rational and integral function of the degree $n-2$; and so on.

Finally we shall have

$$\Delta^n u_x = an(n-1)(n-2)\ldots 1,$$

a constant quantity.

Hence also we have

$$\Delta^n x^n = 1 \cdot 2 \ldots n \ldots\ldots\ldots\ldots\ldots\ldots (2).$$

2. While the operation or series of operations denoted by Δ, Δ^2, ... Δ^n are always possible when the subject function u_x is given, there are certain elementary cases in which the forms of the results are deserving of particular attention, and these we shall next consider.

Differences of Elementary Functions.

1st. Let $u_x = x(x-1)(x-2) \dots (x-m+1)$.

Then by definition,

$$\Delta u_x = (x+1)x(x-1)\dots(x-m+2) - x(x-1)(x-2)\dots(x-m+1)$$
$$= mx(x-1)(x-2)\dots(x-m+2).$$

When the factors of a continued product increase or decrease by a constant difference, or when they are similar functions of a variable which, in passing from one to the other, increases or decreases by a constant difference, as in the expression

$$\sin x \sin(x+h) \sin(x+2h) \dots \sin\{x+(m-1)h\},$$

the factors are usually called *factorials,* and the term in which they are involved is called a factorial term. For the particular kind of factorials illustrated in the above example it is common to employ the notation

$$x(x-1)\dots(x-m+1) = x^{(m)} \dots\dots\dots\dots\dots (1),$$

doing which, we have

$$\Delta x^{(m)} = mx^{(m-1)} \dots\dots\dots\dots\dots (2).$$

Hence, $x^{(m-1)}$ being also a factorial term,

$$\Delta^2 x^{(m)} = m(m-1)x^{(m-2)},$$

and generally

$$\Delta^n x^{(m)} = m(m-1)\dots(m-n+1)x^{(m-n)} \dots\dots\dots (3).$$

2ndly. Let $u_x = \dfrac{1}{x(x+1)\dots(x+m-1)}$.

Then by definition,

$$\Delta u_x = \frac{1}{(x+1)(x+2)\dots(x+m)} - \frac{1}{x(x+1)\dots(x+m-1)}$$

$$= \left(\frac{1}{x+m} - \frac{1}{x}\right) \frac{1}{(x+1)(x+2)\dots(x+m-1)}$$

$$= \frac{-m}{x(x+1)\dots(x+m)} \dots\dots\dots\dots\dots\dots (4).$$

Hence adopting the notation

$$\frac{1}{x\,(x+1)\,\dots\,(x+m-1)} = x^{(-m)},$$

we have

$$\Delta x^{(-m)} = -\,m x^{(-m-1)} \dots\dots\dots\dots\dots (5).$$

Hence by successive repetitions of the operation Δ,

$$\Delta^n x^{(-m)} = -\,m\,(-m-1)\,\dots\,(-m-n+1)\,x^{(-m-n)}$$
$$= (-1)^n\,m\,(m+1)\,\dots\,(m+n-1)\,x^{(-m-n)} \dots\dots (6),$$

and this may be regarded as an extension of (3).

3rdly. Employing the most general form of factorials, we find

$$\Delta u_x u_{x-1} \dots u_{x-m+1} = (u_{x+1} - u_{x-m+1}) \times u_x u_{x-1} \dots u_{x-m+2} \dots\dots (7),$$

$$\Delta \frac{1}{u_x u_{x+1} \dots u_{x+m-1}} = \frac{u_x - u_{x+m}}{u_x u_{x+1} \dots u_{x+2m}} \dots\dots\dots (8),$$

and in particular if $u_x = ax + b$,

$$\Delta u_x u_{x-1} \dots u_{x-m+1} = am\,u_x u_{x-1} \dots u_{x-m+2} \dots\dots\dots (9),$$

$$\Delta \frac{1}{u_x u_{x+1} \dots u_{x+m-1}} = \frac{-am}{u_x u_{x+1} \dots u_{x+m}} \dots\dots\dots (10).$$

In like manner we have

$$\Delta \log u_x = \log u_{x+1} - \log u_x = \log \frac{u_{x+1}}{u_x}.$$

To this result we may give the form

$$\Delta \log u_x = \log \left(1 + \frac{\Delta u_x}{u_x} \right) \dots\dots\dots\dots (11).$$

So also

$$\Delta \log (u_x u_{x-1} \dots u_{x-m+1}) = \log \frac{u_{x+1}}{u_{x-m+1}} \dots\dots\dots (12).$$

4thly. To find the successive differences of a^x.

We have

$$\Delta a^x = a^{x+1} - a^x$$

$$= (a-1) a^x \dots\dots\dots\dots (13).$$

Hence

$$\Delta^2 a^x = (a-1)^2 a^x,$$

and generally,

$$\Delta^n a^x = (a-1)^n a^x \dots\dots\dots\dots(14).$$

Hence also, since $a^{mx} = (a^m)^x$, we have

$$\Delta^n a^{mx} = (a^m - 1)^n a^{mx} \dots\dots\dots\dots (15).$$

5thly.   To deduce the successive differences of $\sin(ax+b)$ and $\cos(ax+b)$.

$$\Delta \sin(ax+b) = \sin(ax+b+a) - \sin(ax+b)$$

$$= 2 \sin \frac{a}{2} \cos \left(ax + b + \frac{a}{2} \right)$$

$$= 2 \sin \frac{a}{2} \sin \left(ax + b + \frac{a+\pi}{2} \right).$$

By inspection of the form of this result we see that

$$\Delta^2 \sin(ax+b) = \left(2 \sin \frac{a}{2} \right)^2 \sin(ax+b+a+\pi) \dots\dots(16).$$

And generally,

$$\Delta^n \sin(ax+b) = \left(2 \sin \frac{a}{2} \right)^n \sin \left\{ ax + b + \frac{n(a+\pi)}{2} \right\} \dots(17).$$

In the same way it will be found that

$$\Delta^n \cos(ax+b) = \left(2 \sin \frac{a}{2} \right)^n \cos \left\{ ax + b + \frac{n(a+\pi)}{2} \right\} \dots(18).$$

These results might also be deduced by substituting for the sines and cosines their exponential values and applying (15).

3.   The above are the most important forms.   The following are added merely for the sake of exercise.

To find the differences of $\tan u_x$ and of $\tan^{-1} u_x$.

$$\Delta \tan u_x = \tan u_{x+1} - \tan u_x$$

$$= \frac{\sin u_{x+1}}{\cos u_{x+1}} - \frac{\sin u_x}{\cos u_x}$$

$$= \frac{\sin (u_{x+1} - u_x)}{\cos u_{x+1} \cos u_x}$$

$$= \frac{\sin \Delta u_x}{\cos u_{x+1} \cos u_x} \quad \dots\dots\dots\dots\dots (1).$$

Next,

$$\Delta \tan^{-1} u_x = \tan^{-1} u_{x+1} - \tan^{-1} u_x$$

$$= \tan^{-1} \frac{u_{x+1} - u_x}{1 + u_{x+1} u_x}$$

$$= \tan^{-1} \frac{\Delta u_x}{1 + u_{x+1} u_x} \quad \dots\dots\dots\dots (2).$$

From the above, or independently, it is easily shewn that

$$\Delta \tan ax = \frac{\sin a}{\cos ax \cos a \, (x + 1)} \quad \dots\dots\dots (3),$$

$$\Delta \tan^{-1} ax = \tan^{-1} \frac{a}{1 + a^2 x + a^2 x^2} \quad \dots\dots\dots (4).$$

Additional examples will be found in the exercises at the end of this chapter.

4.   When the increment of x is indeterminate, the operation denoted by $\frac{\Delta}{\Delta x}$ merges, on supposing Δx to become infinitesimal but the subject function to remain unchanged, into the operation denoted by $\frac{d}{dx}$. The following are illustrations of the mode in which some of the general theorems of the Calculus of Finite Differences thus merge into theorems of the Differential Calculus.

Ex. We have

$$\frac{\Delta \sin x}{\Delta x} = \frac{\sin (x + \Delta x) - \sin x}{\Delta x}$$

$$= \frac{2 \sin \frac{1}{2} \Delta x \, \sin \left(x + \dfrac{\Delta x + \pi}{2} \right)}{\Delta x} .$$

And, repeating the operation n times,

$$\frac{\Delta^n \sin x}{(\Delta x)^n} = \frac{(2 \sin \frac{1}{2} \Delta x)^n \sin \left(x + n \dfrac{\Delta x + \pi}{2} \right)}{(\Delta x)^n} \ \ldots\ldots (1).$$

It is easy to see that the limiting form of this equation is

$$\frac{d^n \sin x}{dx^n} = \sin \left(x + \frac{n\pi}{2} \right) \ \ldots\ldots\ldots\ldots (2),$$

a known theorem of the Differential Calculus.

Again, we have

$$\frac{\Delta a^x}{\Delta x} = \frac{a^{x + \Delta x} - a^x}{\Delta x}$$

$$= \left(\frac{a^{\Delta x} - 1}{\Delta x} \right) a^x .$$

And hence, generally,

$$\frac{\Delta^n a^x}{(\Delta x)^n} = \left(\frac{a^{\Delta x} - 1}{\Delta x} \right)^n a^x \ \ldots\ldots\ldots\ldots (3).$$

Supposing Δx to become infinitesimal, this gives by the ordinary rule for vanishing fractions

$$\frac{d^n a^x}{dx^n} = (\log a)^n a^x \ \ldots\ldots\ldots\ldots (4).$$

But it is not from examples like these to be inferred that the Differential Calculus is merely a particular case of the Calculus of Finite Differences. The true nature of their connexion will be developed in a future chapter (Chap. VIII.).

Expansion by factorials.

5. Attention has been directed to the formal analogy between the differences of factorials and the differential coefficients of powers. This analogy is further developed in the following proposition.

To develope $\phi(x)$, a supposed rational and integral function of x of the mth degree, in a series of factorials.

1st. Assume

$$\phi(x) = a + bx + cx^{(2)} + dx^{(3)} \ldots + hx^{(m)} \ldots\ldots\ldots (1).$$

The legitimacy of this form is evident, for it represents a rational and integral function of x of the nth degree, containing a number of arbitrary coefficients equal to the number of coefficients supposed given in $\phi(x)$. And the actual values of the former might be determined by expressing both members of the equation in ascending powers of x, equating coefficients and solving the linear equations which result. Instead of doing this, let us take the successive differences of (1). We find by (2), Art. 2,

$$\Delta\phi(x) = b + 2cx + 3dx^{(2)} \ldots + mhx^{(m-1)} \ldots\ldots\ldots (2),$$

$$\Delta^2\phi(x) = 2c + 3 \cdot 2dx \ldots + m(m-1)\,hx^{(m-2)} \ldots (3),$$

$$\ldots\ldots\ldots\ldots\ldots\ldots\ldots\ldots\ldots\ldots\ldots\ldots\ldots\ldots\ldots\ldots$$

$$\Delta^m\phi(x) = m\,(m-1) \ldots 1h \ldots\ldots\ldots\ldots\ldots (4).$$

And now making $x = 0$ in the series of equations $(1)\ldots(4)$, and representing by $\Delta\phi(0)$, $\Delta^2\phi(0)$, &c. what $\Delta\phi(x)$, $\Delta^2\phi(x)$, &c. become when $x = 0$, we have

$$\phi(0) = a, \quad \Delta\phi(0) = b, \quad \Delta^2\phi(0) = 2c,$$

$$\ldots\ldots\ldots\ldots\ldots\ldots\ldots\ldots\ldots$$

$$\Delta^m\phi(0) = 1 \cdot 2 \ldots mh.$$

Whence determining $a, b, c, \ldots h$, we have

$$\phi(x) = \phi(0) + \Delta\phi(0)\,x + \frac{\Delta^2\phi(0)}{2}\,x^{(2)} + \frac{\Delta^3\phi(0)}{2 \cdot 3}\,x^{(3)} + \&c. \quad (5).$$

If with greater generality we assume

$$\phi(x) = a + bx + cx\,(x-h) + dx\,(x-h)\,(x-2h) + \&c.,$$

we shall find by proceeding as before, except in the employ-ing of $\dfrac{\Delta}{\Delta x}$ for Δ, where $\Delta x = h$,

$$\phi(x) = \{\phi(x)\} + \left\{\frac{\Delta\phi(x)}{\Delta x}\right\}x + \left\{\frac{\Delta^2\phi(x)}{(\Delta x)^2}\right\}\frac{x(x-h)}{1.2}$$

$$+ \left\{\frac{\Delta^3\phi(x)}{(\Delta x)^3}\right\}\frac{x(x-h)(x-2h)}{1.2.3} + \&\text{c.} \dots (6),$$

where the brackets $\{\ \}$ denote that in the enclosed function, after reduction, x is to be made equal to 0.

Taylor's theorem is the limiting form to which the above theorem approaches when the increment Δx is indefinitely diminished.

General theorems expressing relations between the successive values, successive differences, and successive differential coefficients of functions.

6.   In the equation of definition

$$\Delta u_x = u_{x+1} - u_x$$

we have the fundamental relation connecting the first differ-ence of a function with two successive values of that function. In Taylor's Theorem, expressed in the form

$$u_{x+1} - u_x = \frac{du_x}{dx} + \frac{1}{2}\frac{d^2u_x}{dx^2} + \frac{1}{2.3}\frac{d^3u_x}{dx^3} + \&\text{c.},$$

we see the fundamental relation connecting the first difference of a function with its successive differential coefficients. From these fundamental relations spring many general theo-rems expressing derived relations between the differences of the higher orders, the successive values, and the differential coefficients of functions.

As concerns the history of such theorems it may be ob-served that they appear to have been first suggested by par-ticular instances, and then established, either by that kind of proof which consists in shewing that if a theorem is true for any particular integer value of an index n, it is true for the next greater value, and therefore for all succeeding values; or else by a peculiar method, hereafter to be explained, called the method of Generating Functions. But having been once established, the very forms of the theorems led to a deeper conception of their real nature, and it came to be understood that they were consequences of the formal laws of combination of those operations by which from a given function its succeeding values, its differences, and its differential coefficients are derived.

7. These progressive methods will be illustrated in the following example.

Ex. Required to express u_{x+n} in terms of u_x and its successive differences.

We have

$$u_{x+1} = u_x + \Delta u_x;$$
$$\therefore\ u_{x+2} = u_x + \Delta u_x + \Delta (u_x + \Delta u_x)$$
$$= u_x + 2\Delta u_x + \Delta^2 u_x.$$

Hence proceeding as before we find

$$u_{x+3} = u_x + 3\Delta u_x + 3\Delta^2 u_x + \Delta^3 u_x.$$

These special results suggest, by the agreement of their coefficients with those of the successive powers of a binomial, the general theorem

$$u_{x+n} = u_x + n\Delta u_x + \frac{n(n-1)}{1.2}\Delta^2 u_x$$
$$+ \frac{n(n-1)(n-2)}{1.2.3}\Delta^3 u_x + \&c.\dots\dots(1).$$

Suppose then this theorem true for a particular value of n, then for the next greater value we have

$$u_{x+n+1} = u_x + n\Delta u_x + \frac{n(n-1)}{1.2}\Delta^2 u_x$$
$$+ \frac{n(n-1)(n-2)}{1.2.3}\Delta^3 u_x + \&c.$$
$$+ \Delta u_x + n\Delta^2 u_x + \frac{n(n-1)}{1.2}\Delta^3 u_x + \&c.$$

$$= u_x + (n+1)\,\Delta u_x + \frac{(n+1)\,n}{1\,.\,2}\,\Delta^2 u_x + \frac{(n+1)\,n\,(n-1)}{1\,.\,2\,.\,3}\,\Delta^3 u_x + \&c.$$

the form of which shews that the theorem remains true for the next greater value of n, therefore for the value of n still succeeding, and so on *ad infinitum*. But it is true for $n=1$, and therefore for all positive integer values of n whatever.

8. We proceed to demonstrate the same theorem by the method of generating functions.

Definition. If $\phi\,(t)$ is capable of being developed in a series of powers of t, the general term of the expansion being represented by $u_x t^x$, then $\phi\,(t)$ is said to be the generating function of u_x. And this relation is expressed in the form

$$\phi\,(t) = Gu_x.$$

Thus we have

$$e^t = G\,\frac{1}{1\,.\,2\ldots x},$$

since $\dfrac{1}{1\,.\,2\ldots x}$ is the coefficient of t^x in the development of e^t.

In like manner

$$\frac{e^t}{t} = G\,\frac{1}{1\,.\,2\ldots(x+1)},$$

since $\dfrac{1}{1\,.\,2\ldots(x+1)}$ is the coefficient of t^x in the development of the first member.

And generally, if $Gu_x = \phi\,(t)$, then

$$Gu_{x+1} = \frac{\phi\,(t)}{t}\;\ldots\ldots\; Gu_{x+n} = \frac{\phi\,(t)}{t^n}\;\ldots\ldots\ldots\ldots\; (2).$$

Hence therefore

$$Gu_{x+1} - Gu_x = \left(\frac{1}{t} - 1\right)\phi\,(t).$$

But the first member is obviously equal to $G\Delta u_x$, therefore

$$G\Delta u_x = \left(\frac{1}{t} - 1\right)\phi\,(t)\ldots\ldots\ldots\ldots\ldots (3).$$

And generally

$$G\Delta^n u_x = \left(\frac{1}{t} - 1\right)^n \phi(t) \quad \dotsc\dotsc\dotsc\dotsc (4).$$

To apply these theorems to the problem under consideration we have, supposing still $Gu_x = \phi(t)$,

$$Gu_{x+n} = \left(\frac{1}{t}\right)^n \phi(t)$$

$$= \left\{1 + \left(\frac{1}{t} - 1\right)\right\}^n \phi(t)$$

$$= \phi(t) + n\left(\frac{1}{t} - 1\right)\phi(t) + \frac{n(n-1)}{1.2}\left(\frac{1}{t} - 1\right)^2 \phi(t) + \&c.$$

$$= Gu_x + nG\Delta u_x + \frac{n(n-1)}{2} G\Delta^2 u_x + \&c.$$

$$= G\left\{u_x + n\Delta u_x + \frac{n(n-1)}{2}\Delta^2 u_x + \&c.\right\}$$

Hence

$$u_{x+n} = u_x + n\Delta u_x + \frac{n(n-1)}{2}\Delta^2 u_x + \&c.$$

which agrees with (1).

Although on account of the extensive use which has been
made of the method of generating functions, especially by
the older analysts, we have thought it right to illustrate its
general principles, it is proper to notice that there exists an
objection in point of scientific order to the employment of
the method for the demonstration of the direct theorems of
the Calculus of Finite Difference; viz. that G is, from its
very nature, a symbol of inversion (*Diff. Equations*, p. 375).
In applying it, we do not perform a direct and definite operation, but seek the answer to a question, viz. What is that
function which, on performing the direct operation of development, produces terms possessing coefficients of a certain
form? and this is a question which admits of an infinite
variety of answers according to the extent of the development
and the kind of indices supposed admissible. Hence the
distributive property of the symbol G, as virtually employed

in the above example, supposes limitations which are not implied in the mere definition of the symbol. It must be supposed to have reference to the same system of indices in the one member as in the other; and though, such conventions being supplied, it becomes a strict method of proof, its indirect character still remains.

9. We proceed to the last of the methods referred to in Art 6, viz. that which is founded upon the study of the ultimate laws of the operations involved. In addition to the symbol Δ, we shall introduce a symbol D to denote the operation of giving to x in a proposed subject function the increment unity;—its definition being

$$Du_x = u_{x+1} \quad \dots\dots\dots\dots\dots\dots(1).$$

Laws and Relations of the symbols D, Δ and $\dfrac{d}{dx}$.

1st. The symbol Δ is *distributive* in its operation. Thus

$$\Delta (u_x + v_x + \&c.) = \Delta u_x + \Delta v_x \dots\dots\dots (2).$$

For

$$\Delta (u_x + v_x + \&c.) = u_{x+1} + v_{x+1} \dots - (u_x + v_x \dots)$$

$$= u_{x+1} - u_x + v_{x+1} - v_x \dots$$

$$= \Delta u_x + \Delta v_x \dots$$

In like manner we have

$$\Delta (u_x - v_x \dots) = \Delta u_x - \Delta v_x \dots\dots\dots\dots (3).$$

2ndly. The symbol Δ is *commutative* with respect to any constant coefficients in the terms of the subject to which it is applied. Thus a being constant,

$$\Delta a u_x = a u_{x+1} - a u_x$$

$$= a \Delta u_x \dots\dots\dots\dots\dots (4).$$

And from this law in combination with the preceding one, we have, a, b,... being constants,

$$\Delta (a u_x + b v_x \dots) = a \Delta u_x + b \Delta v_x \dots\dots\dots (5).$$

3rdly. The symbol Δ obeys the index law expressed by the equation

$$\Delta^m \Delta^n u_x = \Delta^{m+n} u_x \dots\dots\dots\dots\dots(6),$$

m and n being positive indices. For, by the implied definition of the index m,

$$\Delta^m \Delta^n u_x = (\Delta\Delta \dots m \text{ times}) (\Delta\Delta \dots n \text{ times}) u_x$$
$$= \{\Delta\Delta \dots (m + n) \text{ times}\} u_x$$
$$= \Delta^{m+n} u_x.$$

These are the primary laws of combination of the symbol Δ. It will be seen from these that Δ combines with Δ and with constant quantities, as symbols of quantity combine with each other. Thus, $(\Delta + a) u$ denoting $\Delta u + au$, we should have, in virtue of the first two of the above laws,

$$(\Delta + a)(\Delta + b) u = \{\Delta^2 + (a + b) \Delta + ab\} u$$
$$= \Delta^2 u + (a + b) \Delta u + abu \dots\dots\dots\dots(7),$$

the developed result of the combination $(\Delta + a)(\Delta + b)$ being in form the same as if Δ were a symbol of quantity.

The index law (6) is virtually an expression of the formal consequences of the truth that Δ denotes an operation which, performed upon any function of x, converts it into another function of x upon which *the same operation may be repeated*. Perhaps it might with propriety be termed the law of repetition;—as such it is common to all symbols of operation, except such, if such there be, as so alter the nature of the subject to which they are applied, as to be incapable of repetition. It was however necessary that it should be distinctly noticed, because it constitutes a part of the formal ground of the general theorems of the calculus.

The laws which have been established for the symbol Δ are even more obviously true for the symbol D. The two symbols are connected by the equation

$$D = 1 + \Delta,$$

since

$$Du_x = u_x + \Delta u_x = (1 + \Delta) u_x \dots\dots\dots\dots(8),$$

and they are connected with $\dfrac{d}{dx}$ by the relation

$$D = \epsilon^{\frac{d}{dx}} \dots\dots\dots\dots\dots\dots (9),$$

founded on the symbolical form of Taylor's theorem.   For

$$Du_x = u_{x+1} = u_x + \frac{d}{dx}u_x + \frac{1}{2}\frac{d^2 u_x}{dx^2} + \frac{1}{2.3}\frac{d^3 u_x}{dx^3} + \&c.$$

$$= \left(1 + \frac{d}{dx} + \frac{1}{2}\frac{d^2}{dx^2} + \frac{1}{2.3}\frac{d^3}{dx^3} + \&c.\right)u_x$$

$$= \epsilon^{\frac{d}{dx}} u_x.$$

It thus appears that D, Δ, and $\dfrac{d}{dx}$, are connected by the two equations

$$D = 1 + \Delta = \epsilon^{\frac{d}{dx}} \dots\dots\dots\dots\dots (10),$$

and from the fact that D and Δ are thus both expressible by means of $\dfrac{d}{dx}$ it might be inferred that the symbols D, Δ, and $\dfrac{d}{dx}$ combine, each with itself, with constant quantities, and with each other, as if they were individually symbols of quantity.   (*Differential Equations*, Chapter XVI.)

10.   In the following section these principles will be applied to the demonstration of what may be termed the direct general theorems of the Calculus of Differences.   The conditions of their inversion, i.e. of their extension to cases in which symbols of operation occur under negative indices, will be considered, so far as may be necessary, in subsequent chapters.

Ex. 1.   To develope u_{x+n} in a series consisting of u_x and its successive differences (Ex. of Art. 7, resumed).

By definition

$$u_{x+1} = Du_x, \quad u_{x+2} = D^2 u_x, \quad \&c.$$

Therefore

$$u_{x+n} = D^n u_x = (1 + \Delta)^n u_x \dots\dots\dots\dots\dots\dots\dots\dots (1),$$

$$= \left\{ 1 + n\Delta + \frac{n(n-1)}{2} \Delta^2 + \frac{n(n-1)(n-2)}{2.3} \Delta^3 \dots \right\} u_x$$

$$= u_x + n\Delta u_x + \frac{n(n-1)}{2} \Delta^2 u_x + \frac{n(n-1)(n-2)}{2.3} \Delta^3 u_x + \&c.. (2).$$

Ex. 2. To express $\Delta^n u_x$ in terms of u_x and its successive values.

Since $\quad \Delta u_x = u_{x+1} - u_x = D u_x - u_x$, we have

$$\Delta u_x = (D - 1) u_x,$$

and as, the operations being performed, each side remains a function of x,

$$\Delta^n u_x = (D - 1)^n u_x$$

$$= \left\{ D^n - n D^{n-1} + \frac{n(n-1)}{1.2} D^{n-2} + \&c. \right\} u_x.$$

Hence, interpreting the successive terms,

$$\Delta^n u_x = u_{x+n} - n u_{x+n-1} + \frac{n(n-1)}{1.2} u_{x+n-2} \dots + (-1)^n u_x \dots..(3).$$

Of particular applications of this theorem those are the most important which result from supposing $u_x = x^m$.

We have

$$\Delta^n x^m = (x+n)^m - n(x+n-1)^m + \frac{n(n-1)}{1.2} (x+n-2)^m - \&c...(4).$$

Now let the notation $\Delta^n 0^m$ be adopted to express what the first member of the above equation becomes when $x = 0$; then

$$\Delta^n 0^m = n^m - n (n-1)^m$$

$$+ \frac{n(n-1)(n-2)^m}{1.2} - \frac{n(n-1)(n-2)(n-3)^m}{1.2.3} + \&c......(5).$$

The systems of numbers expressed by $\Delta^n 0^m$ are of frequent occurrence in the theory of series. Professor De Morgan has

2—2

given the following table of their values up to $\Delta^{10}0^{10}$ calculated from the above theorem. (*Diff. Calculus*, p. 253):

	Δ	Δ^2	Δ^3	Δ^4	Δ^5	Δ^6	Δ^7	Δ^8	Δ^9	Δ^{10}
0	1	0	0	0	0	0	0	0	0	0
0^2	1	2	0	0	0	0	0	0	0	0
0^3	1	6	6	0	0	0	0	0	0	0
0^4	1	14	36	24	0	0	0	0	0	0
0^5	1	30	150	240	120	0	0	0	0	0
0^6	1	62	540	1560	1800	720	0	0	0	0
0^7	1	126	1806	8400	16800	15120	5040	0	0	0
0^8	1	254	5796	40824	126000	191520	141120	40320	0	0
0^9	1	510	18150	186480	834120	1905120	2328480	1451520	362880	0
0^{10}	1	1022	55980	818520	5103000	16435440	29635200	30240000	16329600	3628800

From (2) Art. 1, we have

$$\Delta^n 0^n = 1.2 \ldots n,$$

and, equating this with the corresponding value given by (5), we have

$$1.2 \ldots n = n^n - n(n-1)^n + \frac{n(n-1)}{1.2}(n-2)^n + \&c.\ldots\ldots(6).$$

Ex. 3.   To obtain developed expressions for the nth difference of the product of two functions u_x and v_x.

Since

$$\Delta u_x v_x = u_{x+1} \cdot v_{x+1} - u_x v_x$$
$$= Du_x . D'v_x - u_x v_x,$$

where D applies to u_x alone, and D' to v_x alone, we have

$$\Delta u_x v_x = (DD' - 1) u_x v_x,$$

and generally

$$\Delta^n u_x v_x = (DD' - 1)^n u_x v_x \ldots\ldots\ldots\ldots\ldots(7).$$

It now only remains to transform, if needful, and to develope the operative function in the second member according to the nature of the expansion required.

Thus if it be required to express $\Delta^n u_x v_x$ in ascending differences of v_x, we must change D' into $\Delta' + 1$, regarding Δ' as operating only on v_x. We then have

$$\Delta^n u_x v_x = \{D(1 + \Delta') - 1\}^n u_x v_x$$
$$= (\Delta + D\Delta')^n u_x v_x$$
$$= \left\{\Delta^n + n\Delta^{n-1}D\Delta' + \frac{n(n-1)}{1.2}\Delta^{n-2}D^2\Delta'^2 + \&c.\right\} u_x v_x.$$

Remembering then that Δ and D operate only on u_x and Δ' only on v_x, and that the accent on the latter symbol may be dropped when that symbol only precedes v_x, we have

$$\Delta^n u_x v_x = \Delta^n u_x . v_x + n\Delta^{n-1}u_{x+1} . \Delta v_x$$
$$+ \frac{n(n-1)}{1.2}\Delta^{n-2}u_{x+2} . \Delta^2 v_x + \&c. \ldots\ldots(8),$$

the expansion required.

As a particular illustration, suppose $u_x = a^x$. Then since

$$\Delta^{n-r}u_{x+r} = \Delta^{n-r}a^{x+r} = a^r\Delta^{n-r}a^x$$
$$= a^{x+r}(a-1)^{n-r}, \text{ by (14), Art. 2,}$$

we have

$$\Delta^n a^x v_x = a^x\{(a-1)^n v_x + n(a-1)^{n-1}a\Delta v_x$$
$$+ \frac{n(n-1)}{2}(a-1)^{n-2}a^2\Delta^2 v_x + \&c.\} \ldots\ldots(9).$$

Again, if the expansion is to be ordered according to successive values of v_x, it is necessary to expand the untransformed operative function in the second member of (7) in ascending powers of D' and develope the result. We find

$$\Delta^n u_x v_x = (-1)^n\{u_x v_x - nu_{x+1}v_{x+1} + \frac{n(n-1)}{2}u_{x+2}v_{x+2} + \&c.\}\ldots(10).$$

Lastly, if the expansion is to involve only the differences of u_x and v_x, then, changing D into $1 + \Delta$, and D' into $1 + \Delta'$, we have

$$\Delta^n u_x v_x = (\Delta + \Delta' + \Delta\Delta')^n u_x v_x. \ldots\ldots\ldots\ldots(11),$$

and the symbolic trinomial in the second member is now to be developed and the result interpreted.

Ex. 4. To express $\Delta^n u_x$ in terms of the differential coefficients of u_x.

1st. By (10), Art. 9, $\Delta = \epsilon^{\frac{d}{dx}} - 1$. Hence

$$\Delta^n u_x = (\epsilon^{\frac{d}{dx}} - 1)^n u_x \ldots\ldots\ldots\ldots\ldots(12).$$

Now t being a symbol of quantity, we have

$$(\epsilon^t - 1)^n = \left(t + \frac{t^2}{1.2} + \frac{t^3}{1.2.3} + \&c.\right)^n \ldots\ldots\ldots (13),$$

$$= t^n + A_1 t^{n+1} + A_2 t^{n+2} + \&c.,$$

on expansion, A_1, A_2, &c. being numerical coefficients. Hence

$$(\epsilon^{\frac{d}{dx}} - 1)^n = \left(\frac{d}{dx}\right)^n + A_1 \left(\frac{d}{dx}\right)^{n+1} + A_2 \left(\frac{d}{dx}\right)^{n+2} + \&c.,$$

and therefore

$$\Delta^n u_x = \left(\frac{d}{dx}\right)^n u_x + A_1 \left(\frac{d}{dx}\right)^{n+1} u_x + A_2 \left(\frac{d}{dx}\right)^{n+2} u_x + \&c.\ldots(14).$$

11.   The coefficients A_1, A_2, ...&c. may be determined in various ways, the simplest in principle being perhaps to develope the right-hand member of (13) by the polynomial theorem, and then seek the aggregate coefficients of the successive powers of t.   But the expansion may also be effected with complete determination of the constants by a remarkable secondary form of Maclaurin's theorem, which we shall proceed to demonstrate.

Secondary form of Maclaurin's Theorem.

PROP.   *The development of $\phi(t)$ in positive and integral powers of t, when such development is possible, may be expressed in the form*

$$\phi(t) = \phi(0) + \phi\left(\frac{d}{d0}\right) 0 . t + \phi\left(\frac{d}{d0}\right) 0^2 . \frac{t^2}{1.2}$$

$$+ \phi\left(\frac{d}{d0}\right) 0^3 . \frac{t^3}{1.2.3} + \&c.$$

where $\phi\left(\frac{d}{d0}\right) 0^m$ *denotes what* $\phi\left(\frac{d}{dt}\right) t^m$ *becomes when* $t = 0$.

First, we shall shew that if $\phi(t)$ and $\psi(t)$ are any two functions admitting of development in the form $a + bt + ct^2 +$ &c.,

then $$\phi\left(\frac{d}{dt}\right)\psi(t) = \psi\left(\frac{d}{dt}\right)\phi(t)\ldots\ldots\ldots\ldots(15),$$

provided that t be made equal to 0, after the implied operations are performed.

For, developing all the functions, each member of the above equation is resolved into a series of terms of the form $A\left(\frac{d}{dt}\right)^m t^n$, while in corresponding terms of the two members the order of the indices m and n will be reversed.

Now $\left(\frac{d}{dt}\right)^m t^n$ is equal to 0 if m is greater than n, to $1.2\ldots n$ if m is equal to n, and again to 0 if m is less than n and at the same time t equal to 0; for in this case t^{n-m} is a factor. Hence if $t = 0$,

$$\left(\frac{d}{dt}\right)^m t^n = \left(\frac{d}{dt}\right)^n t^m,$$

and therefore under the same condition the equation (15) is true, or, adopting the notation above explained,

$$\phi\left(\frac{d}{d0}\right)\psi(0) = \psi\left(\frac{d}{d0}\right)\phi(0) \ldots\ldots\ldots\ldots (16).$$

Now by Maclaurin's theorem in its known form

$$\phi(t) = \phi(0) + \frac{d}{d0}\phi(0) . t + \frac{d^2}{d0^2}\phi(0) . \frac{t^2}{1.2} + \&c.\ldots\ldots(17).$$

Hence, applying the above theorem of reciprocity,

$$\phi(t) = \phi(0) + \phi\left(\frac{d}{d0}\right)0 . t + \phi\left(\frac{d}{d0}\right)0^2 . \frac{t^2}{1.2} + \&c. \ldots(18),$$

the secondary form in question. The two forms of Maclaurin's theorem (17), (18) may with propriety be termed *conjugate*.

As a particular illustration, suppose $\phi(t) = (\epsilon^t - 1)^n$, n being a positive integer. Then observing that

$$\phi(0) = 0,$$

$$\phi\left(\frac{d}{d0}\right)0^m = \left(\epsilon^{\frac{d}{d0}} - 1\right)^n 0^m$$

$$= \Delta^n 0^m,$$

we have

$$(\epsilon^t - 1)^n = \Delta^n 0 \cdot t + \Delta^n 0^2 \cdot \frac{t^2}{1.2} + \Delta^n 0^3 \cdot \frac{t^3}{1.2.3} + \&c.$$

But $\Delta^n 0^m$ is equal to 0 if m is less than n, and to $1.2...n$ if m is equal to n; Art. 1.   Hence

$$(\epsilon^t - 1)^n = t^n + \frac{\Delta^n 0^{n+1}}{1.2...n+1} \cdot t^{n+1} + \frac{\Delta^n 0^{n+2}}{1.2...n+2} \cdot t^{n+2} + \&c...(19).$$

The coefficients of this expansion are now given by the table in Art. 10, or by the theorem from which that table is calculated.

Hence therefore, since $\Delta^n u = (\epsilon^{\frac{d}{dx}} - 1)^n u$, we have

$$\Delta^n u = \frac{d^n u}{dx^n} + \frac{\Delta^n 0^{n+1}}{1.2...(n+1)} \frac{d^{n+1} u}{dx^{n+1}} + \frac{\Delta^n 0^{n+2}}{1.2...(n+2)} \frac{d^{n+2} u}{dx^{n+2}} + \&c. \quad (20),$$

the theorem sought.

As a particular example we have

$$\Delta^n x^m = m(m-1)...(m-n+1) x^{m-n}$$

$$+ \frac{\Delta^n 0^{n+1}}{1.2...n+1} m(m-1)...(m-n) x^{m-n-1}... + \Delta^n 0^m...(21).$$

The reasoning employed in the above investigation proceeds upon the assumption that n is a positive integer. The very important case in which $n = -1$ will be considered in another chapter of this work.

Ex. 5.   *To express $\dfrac{d^n u}{dx^n}$ in terms of the successive differences of u.*

Since $\epsilon^{\frac{d}{dx}} = 1 + \Delta$, we have

$$\frac{d}{dx} = \log (1 + \Delta),$$

therefore $\qquad \left(\frac{d}{dx}\right)^{n} = \left\{\log (1 + \Delta)\right\}^{n} \dots\dots\dots\dots (22),$

and the right-hand member must now be developed in ascending powers of Δ.

In the particular case of $n = 1$, we have

$$\frac{du}{dx} = \Delta u - \frac{\Delta^{2}u}{2} + \frac{\Delta^{3}u}{3} - \frac{\Delta^{4}u}{4} + \&c.\dots\dots (23).$$

As a particular application, suppose $u = x(x-1)\dots(x-n+1)$, then adopting the notation of Art. 2,

$$u = x^{(n)},$$

$$\Delta u = nx^{(n-1)},$$

$$\Delta^{2}u = n(n-1)x^{(n-2)},$$

&c.

Therefore $\dfrac{dx^{(n)}}{dx} = nx^{(n-1)} - \dfrac{n(n-1)}{2} x^{(n-2)}$

$$+ \frac{n(n-1)(n-2)}{3} x^{(n-3)} \&c. \dots\dots (24),$$

a theorem which enables us to express a differential coefficient of a factorial in factorials.

It would be easy, but it is needless, to multiply these general theorems, some of those above given being valuable rather as an illustration of principles than for their intrinsic importance. Other examples will occur in the course of this work.

EXERCISES.

1. Establish the following results:

$$\Delta \cot ax = \frac{-\sin a}{\sin ax \sin a (x+1)},$$

$$\Delta \, 2^x \sin \frac{a}{2^x} = 2^{x+2} \sin \frac{a}{2^{x+1}} \left(\sin \frac{a}{2^{x+2}} \right)^2,$$

$$\Delta \tan \frac{a}{2^x} = - \frac{\tan \dfrac{a}{2^{x+1}}}{\cos \dfrac{a}{2^x}},$$

$$\Delta \cot (2^x a) = \frac{-1}{\sin (2^{x+1} a)}$$

2. Shew that $\Delta \dfrac{u_x}{v_x}$ may be expressed in the form

$$\frac{v_x \Delta u_x - u_x \Delta v_x}{v_x v_{x+1}}.$$

3. If $\Delta x = a$ and $\Delta u_x = u_{x+a} - u_x$, prove that

$$u_{x+n} = u_x + \frac{n}{a} \Delta u_x + \frac{n \, (n-a)}{2 \cdot a^2} \Delta^2 u_x + \frac{n \, (n-a) \, (n-2a)}{2 \cdot 3 \cdot a^3} \Delta^3 u_x + \&c.$$

4. Prove the following theorem, viz.

$$\iint \ldots a^x y \, dx^n = a^x \left\{ \log \, (a + a\Delta) \right\}^{-n} y.$$

What complementary function must be added to the second member?

5. Shew that Sir John Herschell's theorem for the development of functions of exponentials, viz.

$$f(e^t) = f(1) + f(1 + \Delta) \, 0 \cdot t + f(1 + \Delta) \, 0^2 \cdot \frac{t^2}{1 \cdot 2} + \&c.,$$

is a consequence of the Theorem of Art. 11.

6. If $x = e^\theta$, prove that

$$\left(\frac{d}{d\theta} \right)^n = \frac{\Delta 0^n}{1} x \frac{d}{dx} + \frac{\Delta^2 0^n}{1 \cdot 2} x^2 \frac{d^2}{dx^2} + \frac{\Delta^3 0^n}{1 \cdot 2 \cdot 3} x^3 \frac{d^3}{dx^3} + \&c.$$

7. Prove the following theorems:

$$\Delta^n 0^{n+1} = \frac{n \, (n+1)}{2} \Delta^n 0^n.$$

8.   If $\Delta u_{x,y} = u_{x+1,y+1} - u_{x,y}$, and if $\Delta^n u_{x,y}$ be expanded in a series of differential coefficients of u, shew that the general term will be

$$\frac{\Delta^n 0^{p+q}}{\Delta^p 0^p \times \Delta^q 0^q} \frac{d^{p+q} u_{x,y}}{dx^p dy^q}.$$

9.   Prove the following theorem, viz.

$$a_0 + a_1 x + \frac{a_2 x^2}{2} + \frac{a_0 x^3}{2.3} + \frac{a_4 x^4}{2.3.4}$$

$$= \epsilon^x \left(a_0 + x\Delta a_0 + \frac{x^2}{2} \Delta^2 a_0 + \frac{x^3}{2.3} \Delta^3 a_0 + \&c. \right),$$

the symbol Δ being defined by the equation $\Delta a_n = a_{n+1} - a_n$.

10.   What class of series would the above theorem enable us to convert from a slow to a rapid convergence?

11.   Shew by the theorem of Art. 11, that the development of the function ϵ^{ϵ^t} in ascending powers of t may be exhibited in the form

$$\epsilon \left\{ 1 + (\epsilon^\Delta 0) t + (\epsilon^\Delta 0^2) \frac{t^2}{1.2} + (\epsilon^\Delta 0^3) \frac{t^3}{1.2.3} \cdots \right\}.$$

And hence calculate the first four terms of the expansion.

(28)

CHAPTER III.

OF INTERPOLATION.

1. THE word interpolate has been adopted in analysis to denote primarily the interposing of missing terms in a series of quantities supposed subject to a determinate law of magnitude, but secondarily and more generally to denote the calculating, under some hypothesis of law or continuity, of *any* term of a series from the values of any other terms supposed given.

As no series of *particular* values can determine a law, the problem of interpolation is an indeterminate one. To find an analytical expression of a function from a limited number of its numerical values corresponding to given values of its independent variable x is, in Analysis, what in Geometry it would be to draw a continuous curve through a number of given points. And as in the latter case the number of possible curves, so in the former the number of analytical expressions satisfying the given conditions, is infinite. Thus the form of the function—the species of the curve—must be assumed *a priori*. It may be that the evident character of succession in the values observed indicates what kind of assumption is best. If for instance these values are of a periodical character, circular functions ought to be employed. But where no such indications exist it is customary to assume for the general expression of the values under consideration a rational and integral function of x, and to determine the coefficients by the given conditions.

This assumption rests upon the supposition (a supposition however actually verified in the case of all tabulated functions) that the successive orders of differences rapidly diminish. In the case of a rational and integral function of x of the nth degree it has been seen that differences of the $n+1^{th}$

and of all succeeding orders vanish. Hence if in any other function such differences become very small, that function may, quite irrespectively of its form, be approximately represented by a function which is rational and integral. Of course it is supposed that the value of x for which that of the function is required is not very remote from those, or from some of those, values for which the values of the function are given. The same assumption as to the form of the unknown function and the same condition of limitation as to the use of that form flow in an equally obvious manner from the expansion in Taylor's theorem.

2. The problem of interpolation assumes different forms, according as the values given are equidistant, i. e. correspondent to equidifferent values of the independent variable, or not. But the solution of all its cases rests upon the same principle. The most *obvious* mode in which that principle can be applied is the following. If for n values $a, b, \ldots$ of an independent variable x the corresponding values $u_a, u_b, \ldots$ of an unknown function of x represented by u_x, are given, then, assuming as the approximate general expression of u_x,

$$u_x = A + Bx + Cx^2 \ldots + Ex^{n-1} \ldots\ldots\ldots\ldots (1),$$

a form which is rational and integral and involves n arbitrary coefficients, the data in succession give

$$u_a = A + Ba + Ca^2 \ldots + Ea^{n-1},$$
$$u_b = A + Bb + Cb^2 \ldots + Eb^{n-1},$$
$$\ldots\ldots\ldots\ldots\ldots\ldots\ldots\ldots$$

a system of n linear equations which determine $A, B \ldots E$. To avoid the solving of these equations other but equivalent modes of procedure are employed, all such being in effect reducible to the two following, viz. either to an application of that property of the rational and integral function in the second member of (1) which is expressed by the equation $\Delta^n u_x = 0$, or to the substitution of a different but equivalent form for the rational and integral function. These methods will be respectively illustrated in Prop. 1 and its deductions, and in Prop. 2, of the following sections.

PROP. 1. Given n consecutive equidistant values $u_0, u_1 \ldots u_{n-1}$ of a function u_x, to find its approximate general expression.

By Chap. II. Art. 10,

$$u_{x+m} = u_x + m\Delta u_x + \frac{m\,(m-1)}{1.2}\,\Delta^2 u_x + \&c.$$

Hence, substituting 0 for x, and x for m, we have

$$u_x = u_0 + x\Delta u_0 + \frac{x\,(x-1)}{1.2}\,\Delta^2 u_0 + \&c.\ ad\ inf.$$

But on the assumption that the proposed expression is rational and integral and of the degree $n-1$, we have $\Delta^n u_x = 0$, and therefore $\Delta^n u_0 = 0$. Hence

$$u_x = u_0 + x\Delta u_0 + \frac{x\,(x-1)}{1.2}\,\Delta^2 u_0 \ldots$$

$$+ \frac{x\,(x-1)\ldots(x-n+2)}{1.2\ldots(n-1)}\,\Delta^{n-1} u_0 \ldots\ldots\ldots (2),$$

the expression required. It will be observed that the second member is really a rational and integral function of x of the degree $n-1$, while the coefficients are made determinate by the data.

In applying this theorem the value of x may be conceived to express the *distance* of the term sought from the first term in the series, the common distance of the terms given being taken as unity.

Ex. Given log $3.14 = .4969296$, log $3.15 = .4983106$, log $3.16 = .4996871$, log $3.17 = .5010593$; required an approximate value of log 3.14159.

Here, omitting the decimal point, we have the following table of numbers and differences:

	u_0	u_1	u_2	u_3
	4969296	4983106	4996871	5010593
Δ	13810	13765	13722	
Δ^2	−45	−43		
Δ^3	2			

The first column gives the values of u_0 and its differences up to $\Delta^3 u_0$. Now the common difference of 3.14, 3.15, &c.

being taken as unity, the value of x which corresponds to $3 \cdot 14159$ will be $\cdot 159$. Hence we have

$$u = 4969296 + \cdot 159 \times 13810 + \frac{(\cdot 159)\,(\cdot 159 - 1)}{1.2} \times -45$$
$$+ \frac{(\cdot 159)\,(\cdot 159 - 1)\,(\cdot 159 - 2)}{1.2.3}.$$

Effecting the calculations we find $u_x = \cdot 4971495$, which is true to the last place of decimals. Had the first difference only been employed, which is equivalent to the ordinary rule of proportional parts, there would have been an error of 4 in the last decimal, and second differences would have reduced this error to 1.

3. When the values given and that sought constitute a series of equidistant terms, whatever may be the position of the value sought in that series, it is better to proceed as follows.

Let $u_0, u_1, u_2 \ldots u_n$ be the series. Then since, according to the principle of the method, $\Delta^n u_0 = 0$, we have by Chap. II. Art. 10,

$$u_n - n u_{n-1} + \frac{n\,(n-1)}{1.2}\, u_{n-2} \ldots + (-1)^n u_0 = 0 \ldots \ldots (3),$$

an equation from which any one of the quantities

$$u_0,\ u_1 \ldots u_n$$

may be found in terms of the others.

Thus, to interpolate a term midway between two others, we have

$$u_0 - 2u_1 + u_2 = 0;\quad \therefore\ u_1 = \frac{u_0 + u_2}{2} \ldots \ldots \ldots \ldots (4).$$

Here the middle term is only the arithmetical mean.

To supply the middle term in a series of five, we have

$$u_0 - 4u_1 + 6u_2 - 4u_3 + u_4 = 0;$$
$$\therefore\ u_2 = \frac{4\,(u_1 + u_3) - (u_0 + u_4)}{6} \ldots \ldots \ldots \ldots (5).$$

Ex. Representing as is usual $\int_{0}^{\infty} \epsilon^{-\theta}\theta^{n-1} d\theta$ by $\Gamma(n)$, it is required to complete the following table by finding approximately $\log \Gamma\left(\dfrac{1}{2}\right)$:

n	$\log \Gamma(n)$,	n	$\log \Gamma(n)$,
$\dfrac{2}{12}$	·74556,	$\dfrac{7}{12}$	·18432,
$\dfrac{3}{12}$	·55938,	$\dfrac{8}{12}$	·13165,
$\dfrac{4}{12}$	·42796,	$\dfrac{9}{12}$	·08828,
$\dfrac{5}{12}$	·32788,	$\dfrac{10}{12}$	·05261.

Let the series of values of $\log \Gamma(n)$ be represented by $u_1, u_2 \ldots u_9$, the value sought being that of u_5. Then proceeding as before, we find

$$u_1 - 8u_2 + \frac{8 \cdot \dot{7}}{1 \cdot 2} u_3 - \frac{8 \cdot 7 \cdot 6}{1 \cdot 2 \cdot 3} u_4, \; \&c. = 0,$$

or,

$$u_1 + u_9 - 8\,(u_2 + u_8) + 28\,(u_3 + u_7) - 56\,(u_4 + u_6) + 70u_5 = 0 ;$$

whence

$$u_5 = \frac{56\,(u_4 + u_6) - 28\,(u_3 + u_7) + 8\,(u_2 + u_8) - (u_1 + u_9)}{70} \quad \ldots\ldots (6).$$

Substituting for u_1, u_2, &c., their values from the table, we find

$$\log \Gamma\left(\frac{1}{2}\right) = \text{·}24853,$$

the true value being ·24858.

To shew the gradual closing of the approximation as the number of the values given is increased, the following results are added:

Data.					Calculated value of u_5.
u_4	u_6				.... ·25610,
u_3, u_4	u_6, u_7				.... ·24820,
u_2, u_3, u_4	u_6, u_7, u_8				.... ·24865,
u_1, u_2, u_3, u_4	u_6, u_7, u_8, u_9				.... ·24853.

4. By an extension of the same method, we may treat any case in which the terms given and sought are terms, but not consecutive terms, of a series. Thus, if u_1, u_4, u_5 were given and u_3 sought, the equations $\Delta^3 u_1 = 0$, $\Delta^3 u_2 = 0$ would give

$$u_4 - 3u_3 + 3u_2 - u_1 = 0,$$

$$u_5 - 3u_4 + 3u_3 - u_2 = 0,$$

from which, eliminating u_2, we have

$$3u_5 - 8u_4 + 6u_3 - u_1 = 0 \dots\dots\dots\dots (7),$$

and hence u_3 can be found. But it is better to apply at once the general method of the following Proposition.

PROP. 2. Given n values of a function which are not consecutive and equidistant, to find any other value whose place is given.

Let u_a, u_b, $u_c \dots u_n$ be the given values, corresponding to a, b, $c \dots n$ respectively as values of x, and let it be required to determine an approximate general expression for u_x.

We shall assume this expression rational and integral, Art. 1.

Now there being n conditions to be satisfied, viz. that for $x = a$, $x = b \dots x = n$, it shall assume the respective values u_a, $u_b \dots u_n$, the expression must contain n constants, whose values those conditions determine.

We might therefore assume

$$u_x = A + Bx + Cx^2 \dots + Ex^{n-1} \dots\dots\dots\dots (8),$$

and determine A, B, C by the linear system of equations formed by making $x = a$, $b \dots n$, in succession.

B. F. D.   3

The substitution of another but equivalent form for (8) enables us to dispense with the solution of the linear system.

Let
$$u_x = A (x-b) (x-c) \dots (x-n)$$
$$+ B (x-a) (x-c) \dots (x-n)$$
$$+ C (x-a) (x-b) \dots (x-n) \dots\dots\dots\dots (9)$$

+ &c. to n terms, each of the n terms in the right-hand member wanting one of the factors $x-a$, $x-b$, $\dots (x-n)$, and each being affected with an arbitrary constant. The assumption is legitimate, for the expression thus formed is, like that in (8), rational and integral, and it contains n arbitrary coefficients.

Making $x = a$, we have
$$u_a = A (a-b) (a-c) \dots (a-n) ;$$
therefore
$$A = \frac{u_a}{(a-b) (a-c) \dots (a-n)} .$$

In like manner making $x = b$, we have
$$B = \frac{u_b}{(b-a) (b-c) \dots (b-n)} ,$$

and so on. Hence finally,
$$u_x = u_a \frac{(x-b) (x-c) \dots (x-n)}{(a-b) (a-c) \dots (a-n)} + u_b \frac{(x-a) (x-c) \dots (x-n)}{(b-a) (b-c) \dots (b-n)} \dots$$
$$\dots + u_n \frac{(x-a) (x-b) (x-c) \dots}{(n-a) (n-b) (n-c) \dots} \dots\dots\dots\dots (10),$$

the expression required. This is Lagrange's theorem for interpolation.

As the problem of interpolation, under the assumption that the function to be determined is rational and integral, is a determinate one, the different methods of solution above exemplified lead to consistent results. All these methods are implicitly contained in that of Lagrange.

The following are particular applications of Lagrange's theorem.

5. Given any number of values of a magnitude as observed at given times; to determine approximately the values of the successive differential coefficients of that magnitude at another given time.

Let $a, b, \ldots n$ be the times of observation, $u_a, u_b, \ldots u_n$ the observed values, x the time for which the value is required, and u_x that value. Then the value of u_x is given by (10), and the differential coefficients can thence be deduced in the usual way. But it is most convenient to assume the time represented above by x as the epoch, and to regard $a, b, \ldots n$ as measured from that epoch, being negative if measured backward. The values of $\dfrac{du_x}{dx}$, $\dfrac{d^2u_x}{dx^2}$, &c. will then be the coefficients of x, x^2, &c. in the development of the second member of (10) divided by $1, 1.2, 1.2.3$, &c. successively. Their general expressions may thus at once be found. Thus in particular we shall have

$$\frac{du_x}{dx} = \pm \frac{bc \ldots n \left(\dfrac{1}{b} + \dfrac{1}{c} \ldots + \dfrac{1}{n} \right)}{(a-b)(a-c) \ldots (a-n)} u_a + \&c. \ldots \ldots (11),$$

$$\frac{d^2u_x}{dx^2} = \mp \frac{bc \ldots n \left(\dfrac{1}{bc} + \dfrac{1}{bd} + \dfrac{1}{cd} + \&c. \right)}{1.2 \, (a-b)(a-c) \ldots (a-n)} u_a + \&c. \ldots (12).$$

Laplace's computation of the orbit of a comet is founded upon this proposition (*Mécanique Céleste*).

6. The values of a quantity, e. g. the altitude of a star at given times, are found by observation. Required at what intermediate time the quantity had another given value.

Though it is usual to consider the time as the independent variable, in the above problem it is most convenient to consider the observed magnitude as such, and the time as a function of that magnitude. Let then $a, b, c, \ldots$ be the values given by observation, $u_a, u_b, u_c, \ldots$ the corresponding times, x the value for which the time is sought, and u_x that time. Then the value of u_x is given at once by Lagrange's theorem (10).

The problem may however be solved by regarding the time as the independent variable. Representing then, as in the last example, the times given by a, b, ... n, the time sought by x, and the corresponding values of the observed magnitude by u_a, u_b, ... u_n, and u_x, we must by the solution of the same equation (10) determine x.

The above forms of solution being derived from different hypotheses, will of course differ. We say derived from different hypotheses, because whichsoever element is regarded as dependent is treated not simply as a function, but as a rational and integral function of the other element; and thus the choice affects the nature of the connexion. Except for the avoidance of difficulties of solution, the hypothesis which assumes the time as the independent variable is to be preferred.

Ex. Three observations of a quantity near its time of maximum or minimum being taken, to find its time of maximum or minimum.

Let a, b, c, represent the times of observation, and u_x the magnitude of the quantity at any time x. Then u_a, u_b and u_c are given, and, by Lagrange's formula,

$$u_x = u_a \frac{(x-b)(x-c)}{(a-b)(a-c)} + u_b \frac{(x-c)(x-a)}{(b-c)(b-a)} + u_c \frac{(x-a)(x-b)}{(c-a)(c-b)},$$

and this function of x is to be a maximum or minimum. Hence equating to 0 its differential coefficient with respect to x, we find

$$x = \frac{(b^2 - c^2)\,u_a + (c^2 - a^2)\,u_b + (a^2 - b^2)\,u_c}{2\,\{(b-c)\,u_a + (c-a)\,u_b + (a-b)\,u_c\}} \ \cdots\cdots \ (13).$$

This formula enables us to approximate to the meridian altitude of the sun or of a star when a true meridian observation cannot be taken.

Areas of Curves.

7. Formulæ of interpolation may be applied to the approximate evaluation of integrals between given limits, and therefore to the determination of the areas of curves, the con-

tents of solids, &c. The application is convenient, as it does not require the form of the function under the sign of integration to be known.

PROP. The area of a curve being divided into n portions bounded by $n + 1$ equidistant ordinates $u_0, u_1, \ldots u_n$, whose values, together with their common distance, are given, an approximate expression for the area is required.

The general expression for an ordinate being u_x, we have, if the common distance of the ordinates be assumed as the unit of measure, to seek an approximate value of $\int_0^n u_x dx$.

Now, by (2),

$$u_x = u_0 + x\Delta u_0 + \frac{x(x-1)}{1.2}\Delta^2 u_0 + \frac{x(x-1)(x-2)}{1.2.3}\Delta^3 u_0 + \&c.$$

Hence

$$\int_0^n u_x dx = u_0 \int_0^n dx + \Delta u_0 \int_0^n x dx + \frac{\Delta^2 u_0}{1.2}\int_0^n x(x-1)\, dx$$

$$+ \frac{\Delta^3 u_0}{1.2.3}\int_0^n x(x-1)(x-2)\, dx + \&c.,$$

and effecting the integrations

$$\int_0^n u_x dx = nu_0 + \frac{n^2}{2}\Delta u_0 + \left(\frac{n^3}{3} - \frac{n^2}{2}\right)\frac{\Delta^2 u_0}{1.2} + \left(\frac{n^4}{4} - n^3 + n^2\right)\frac{\Delta^3 u_0}{1.2.3}$$

$$+ \left(\frac{n^5}{5} - \frac{3n^4}{2} + \frac{11n^3}{3} - 3n^2\right)\frac{\Delta^4 u_0}{1.2.3.4}$$

$$+ \left(\frac{n^6}{6} - 2n^5 + \frac{35}{4}n^4 - \frac{50}{3}n^3 + 12n^2\right)\frac{\Delta^5 u_0}{1.2.3.4.5}$$

$$+ \left(\frac{n^7}{7} - \frac{15n^6}{6} + 17n^5 - \frac{225n^4}{4} + \frac{274n^3}{3} - 60n^2\right)\frac{\Delta^6 u_0}{1.2\ldots 6}$$

$$+ \&c. \ldots\ldots\ldots\ldots\ldots\ldots\ldots\ldots\ldots (14).$$

It will be observed that the data permit us to calculate the successive differences of u_0 up to $\Delta^n u_0$. Hence, on the assumption that all succeeding differences may be neglected, the above theorem gives an approximate value of the integral sought. The following are particular deductions.

1st. Let $n = 2$. Then, rejecting all terms after the one involving $\Delta^2 u_0$, we have

$$\int_0^2 u_x dx = 2u_0 + 2\Delta u_0 + \tfrac{1}{3}\Delta^2 u_0.$$

But $\Delta u_0 = u_1 - u_0$, $\Delta^2 u_0 = u_2 - 2u_1 + u_0$; whence, substituting and reducing,

$$\int_0^2 u_x dx = \frac{u_0 + 4u_1 + u_2}{3}.$$

If the common distance of the ordinates be represented by h, the theorem obviously becomes

$$\int_0^{2h} u_x dx = \frac{u_0 + 4u_1 + u_2}{3h} \dots\dots\dots\dots(15),$$

and is the foundation of a well-known rule in treatises of Mensuration.

2ndly. If there are four ordinates whose common distance is h, we find in like manner

$$\int_0^{3h} u_x dx = \frac{3(u_0 + 3u_1 + 3u_2 + u_3)}{8h} \dots\dots\dots (16).$$

3rdly. If five equidistant ordinates are given, we have in like manner

$$\int_0^{4h} u_x dx = \frac{14(u_0 + u_4) + 64(u_1 + u_3) + 24u_2}{45h} \dots\dots (17).$$

4thly. The supposition that the area is divided into six portions bounded by 7 equidistant ordinates leads to a remarkable result, first given by the late Mr Weddle (*Math. Journal*, Vol. IX. p. 79), and deserves to be considered in detail.

Supposing the common distance of the ordinates to be unity, we find, on making $n = 7$ in (14) and calculating the coefficients,

$$\int_0^6 u_x dx = 6u_0 + 18\Delta u_0 + 27\Delta^2 u_0 + 24\Delta^3 u_0 + \frac{123}{10}\Delta^4 u_0$$

$$+ \frac{33}{10}\Delta^5 u_0 + \frac{41}{140}\Delta^6 u_0 \ldots\ldots (18).$$

Now the last coefficient $\frac{41}{140}$ differs from $\frac{42}{140}$ or $\frac{3}{10}$ by the small fraction $\frac{1}{140}$, and as from the nature of the approximation we must suppose sixth differences small, since all succeeding differences are to be neglected, we shall commit but a slight error if we change the last term into $\frac{3}{10}\Delta^6 u_0$. Doing this, and then replacing Δu_0 by $u_1 - u_0$ and so on, we find, on reduction,

$$\int_0^6 u_x dx = \frac{3}{10}\{u_0 + u_2 + u_4 + u_6 + 5(u_1 + u_5) + 6u_3\},$$

which, supposing the common distance of the ordinates to be h, gives

$$\int_0^{6h} u_x dx = \frac{3h}{10}\{u_0 + u_2 + u_4 + u_6 + 5(u_1 + u_5) + 6u_3\}\ldots(19),$$

the formula required.

It is remarkable that, were the series in the second member of (18) continued, the coefficient of $\Delta^7 u_0$ would be found to vanish. Thus while the above formula gives the exact area when fifth differences are constant, it errs in excess by only $\frac{1}{140}\Delta^8 u_0$ when seventh differences are constant.

The practical rule hence derived, and which ought to find a place in elementary treatises on mensuration, is the following:

The proposed area being divided into six portions by seven equidistant ordinates, add into one sum the even ordinates 5 times the odd ordinates and the middle ordinate, and multiply the result by $\frac{3}{10}$ of the common distance of the ordinates.

Ex. 1. The two radii which form a diameter of a circle are bisected, and perpendicular ordinates are raised at the points of bisection. Required the area of that portion of the circle which is included between the two ordinates, the diameter, and the curve, the radius being supposed equal to unity.

The values of the seven equidistant ordinates are

$$\frac{\sqrt{3}}{2}, \ \frac{\sqrt{8}}{3}, \ \frac{\sqrt{35}}{6}, \ 1, \ \frac{\sqrt{35}}{6}, \ \frac{\sqrt{8}}{3}, \ \frac{\sqrt{3}}{2},$$

and the common distance of the ordinates is $\frac{1}{6}$. The area hence computed to five places of decimals is ·95661, which, on comparison with the known value $\frac{\pi}{6} + \frac{\sqrt{3}}{2}$, will be found to be correct to the last figure.

The rule for equidistant ordinates commonly employed would give ·95658.

In all these applications it is desirable to avoid extreme differences among the ordinates. Applied to the quadrant of a circle Mr Weddle's rule, though much more accurate than the ordinary one, leads to a result which is correct only to two places of decimals.

Should the function to be integrated become infinite at or within the limits, an appropriate transformation will be needed.

Ex. 2. Required an approximate value of $\int_0^{\frac{\pi}{2}} \log \sin \theta d\theta$.

The function $\log \sin \theta$ becoming infinite at the lower limit, we have, on integrating by parts,

$$\int \log \sin \theta d\theta = \theta \log \sin \theta - \int \theta \cot \theta d\theta,$$

but the integrated term vanishing at both limits,

$$\int_0^{\frac{\pi}{2}} \log \sin \theta d\theta = - \int_0^{\frac{\pi}{2}} \theta \cot \theta d\theta.$$

The values of the function $\theta \cot \theta$ being now calculated for the successive values $\theta = 0$, $\theta = \dfrac{\pi}{12}$, $\theta = \dfrac{2\pi}{12}$, ... $\theta = \dfrac{\pi}{2}$, and the theorem being applied, we find

$$-\int_0^{\frac{\pi}{2}} \theta \cot \theta \, d\theta = -\cdot 69311.$$

The true value of the definite integral is known to be

$$\frac{\pi}{2} \log_e \left(\frac{1}{2} \right), \text{ or } - \cdot 69316.$$

8. Lagrange's formula enables us to avoid the intermediate employment of differences, and to calculate directly the coefficient of u_m in the general expression for $\int u dx$. If we represent the equidistant ordinates, $2n + 1$ in number, by $u_0, u_1 \ldots u_{2n}$, and change the origin of the integrations by assuming $x - n = y$, we find ultimately

$$\int u dx = A_0 u_n + A_1 (u_{n+1} + u_{n-1}) + A_2 (u_{n+2} + u_{n-2}) \ldots + A_n (u_{2n} + u_0),$$

where generally

$$A_r = \frac{(-1)^r}{1.2 \ldots (n+r) \, 1.2 \ldots (n-r)}$$

$$\times \int_{-n}^{n} \frac{(0^2 - y^2)(1^2 - y^2) \ldots (n^2 - y^2)}{r^2 - y^2} dy \ldots \ldots \ldots (20).$$

A similar formula may be established when the number of equidistant ordinates is even.

Application to Statistics.

9. When the results of statistical observations are presented in a tabular form it is sometimes required to narrow the intervals to which they correspond, or to fill up some particular hiatus by the interpolation of intermediate values. In applying to this purpose the methods of the foregoing sections, it is not to be forgotten that the assumptions which they involve render our conclusions the less trustworthy in proportion as the matter of inquiry is less under the dominion of any known laws, and that this is still more the case in proportion as the field of observation is too narrow to exhibit

fairly the operation of the unknown laws which do exist. The anomalies, for instance, which we meet with in the attempt to estimate the law of human mortality seem rather to be due to the imperfection of our data than to want of continuity in the law itself. The following is an example of the anomalies in question.

Ex. The expectation of life at a particular age being defined as the average duration of life after that age, it is required from the following data, derived from the Carlisle tables of mortality, to estimate the probable expectation of life at 50 years, and in particular to shew how that estimate is affected by the number of the data taken into account.

Age.	Expectation.	Age.	Expectation.
10	$48\cdot82 = u_1$	60	$14\cdot34 = u_6$
20	$41\cdot46 = u_2$	70	$9\cdot18 = u_7$
30	$34\cdot34 = u_3$	80	$5\cdot51 = u_8$
40	$27\cdot61 = u_4$	90	$3\cdot28 = u_9$

The expectation of life at 50 would, according to the above scheme, be represented by u_5. Now if we take as our only data the expectation of life at 40 and 60, we find by the method of Art. 3,

$$u_5 = \frac{u_4 + u_6}{2} = 20\cdot97 \quad\ldots\ldots\ldots\ldots\ldots\ldots(a).$$

If we add to our data the expectation at 30 and 70, we find

$$u_5 = \frac{2}{3}(u_4 + u_6) - \frac{1}{6}(u_3 + u_7) = 20\cdot71 \quad\ldots\ldots(b).$$

If we add the further data for 20 and 80, we find

$$u_5 = \frac{3}{4}(u_4 + u_6) - \frac{3}{10}(u_3 + u_7) + \frac{1}{20}(u_2 + u_8) = 20\cdot75\ldots(c).$$

And if we add in the extreme data for the ages of 10 and 90, we have

$$u_5 = \frac{8}{10}(u_4 + u_6) - \frac{4}{10}(u_3 + u_7)$$
$$+ \frac{8}{70}(u_2 + u_8) - \frac{1}{70}(u_1 + u_9) = 20\cdot77\ldots\ldots\ldots(d).$$

We notice that the second of the above results is considerably lower than the first, but that the second, third, and fourth exhibit a gradual approximation toward some value not very remote from 20·8.

Nevertheless the actual expectation at 50 as given in the Carlisle tables is 21·11, which is greater than even the first result or the average between the expectations at 40 and 60. We may almost certainly conclude from this that the Carlisle table errs in excess for the age of 50.

And a comparison with some recent tables shews that this is so. From the tables of the Registrar-General, Mr Neison* deduced the following results.

Age.	Expectation.	Age.	Expectation.
10	47·7564	60	14·5854
20	40·6910	70	9·2176
30	34·0990	80	5·2160
40	27·4760	90	2·8930
50	20·8463		

Here the calculated values of the expectation at 50, corresponding to those given in (a), (b), (c), (d), will be found to be

$$21\cdot0307, \quad 20\cdot8215, \quad 20\cdot8464, \quad 21\cdot2073.$$

We see here that the actual expectation at 50 is less than the mean between those at 40 and 60. We see also that the second result gives a close, and the third a very close, approximation to its value. The deviation in the fourth result, which takes account of the extreme ages of 10 and 90, seems due to the attempt to comprehend under the same law the mortality of childhood and of extreme old age.

When in an extended table of numerical results the differences tend first to diminish and afterwards to increase, and I think I have observed such a disposition in tables of mortality, it may be concluded that the extreme portions of the tables are subject to different laws. And even should those laws admit, as perhaps they always do, of comprehension

* *Contributions to Vital Statistics,* p. 8.

under some law higher and more general, it may be inferred that that law is incapable of approximate expression in the particular form (Art. 2) which our methods of interpolation presuppose.

EXERCISES.

1. Required an approximate value of log 212 from the following data :

log 210 = 2·3222193,   log 213 = 2·3283796,

log 211 = 2·3242825,   log 214 = 2·3304138.

2. Express v_2 and v_3 approximately in terms of v_0, v_1, v_4, v_5.

3. Find a rational and integral function of x which, when $x = 1, 2, 3$, shall assume the respective values 4, 6, 10.

CHAPTER IV.

FINITE INTEGRATION.

1. THE term integration is here used to denote the process by which, from a given proposed function of x, we determine some other function of which the given function expresses the *difference*.

Thus to integrate u_x is to find a function v_x such that

$$\Delta v_x = u_x.$$

The operation of integration is therefore by definition the inverse of the operation denoted by the symbol Δ. As such it may with perfect propriety be denoted by the inverse form Δ^{-1}. It is usual however to employ for this purpose a distinct symbol, Σ, the origin of which, as well as of the term integration by which its office is denoted, it will be proper to explain.

One of the most important applications of the Calculus of Finite Differences is to the finite summation of series.

Now let u_0, u_1, u_2, &c. represent successive terms of a series whose general term is u_x, and let

$$v_x = u_a + u_{a+1} + u_{a+2} \ldots + u_{x-1} \ldots\ldots\ldots\ldots(1).$$

Then, a being constant so that u_a remains the initial term, we have

$$v_{x+1} = u_a + u_{a+1} + \ldots + u_{x-1} + u_x \ldots\ldots\ldots (2).$$

Hence, subtracting (1) from (2)

$$\Delta v_x = u_x, \therefore v_x = \Delta^{-1} u_x.$$

It appears from the last equation that Δ^{-1}, applied to u_x, expresses the sum of that portion of a series whose general term is u_x, which begins with any fixed term u_a and ends with u_{x-1}. On this account Δ^{-1} has been usually replaced by the symbol Σ, considered as indicating a *summation* or *integration*. At the same time the properties of the symbol Σ, and the mode of performing the operation which it denotes, or, to speak with greater strictness, of answering that question of which it is virtually an expression, are best deduced, and are usually deduced, from its definition as the inverse of the symbol Δ.

For although, considering Σu_x as defined by the equation

$$\Sigma u_x = u_{x-1} + u_{x-2} \ldots + u_a \ldots\ldots\ldots\ldots\ldots\ldots\ldots(3),$$

it may be regarded as denoting a direct and always possible operation, yet, considering it as defined by the equation

$$\Sigma u_x = \Delta^{-1} u_x \ldots\ldots\ldots\ldots\ldots\ldots\ldots(4),$$

and as having for its object the discovery of some finite expression of which the right-hand member of (3) constitutes the development, it is interrogative rather than directive (*Differential Equations*, p. 377); it sets before us an object of inquiry, but does not prescribe any mode of attaining that object. And in fact our knowledge of the cases in which Σu_x admits of finite expression rests *ultimately* upon an examination of the known results of the operation Δ, and a more or less direct reference of the form of the function u_x to such results.

Now one conclusion flows immediately from such reference to the effect of the operation Δ. It is that, whatever may be the form of u_x, the complete expression for Σu_x must contain an arbitrary constant, or, to speak more strictly, an arbitrary quantity which does not change when x changes to $x+1$, and which is therefore constant relatively to the kind of change denoted by the symbol Δ. For let v_x be a particular form of Σu_x, and let $v_x + w_x$ be any other form, this assumption being sufficiently general to include all possible forms if w_x is regarded as an arbitrary function of x, then

$$\Delta v_x = u_x, \quad \Delta\,(v_x + w_x) = u_x.$$

Hence

$$\Delta w_x = 0 \ldots\ldots\ldots\ldots\ldots\ldots\ldots(5).$$

And this simply indicates that w_x is a function of x which does not change when x is increased by unity.

Since w_x is thus constant relatively to Δ, we may with propriety represent it by c, and term it a periodical constant. With this convention we are permitted to say; *If v_x be a particular value of Σu_x, the complete value will be $v_x + c$.*

The necessity of a periodical constant c to complete the value of Σu_x may also be established, and its analytical expression determined, by transforming the problem of summation into that of the solution of a differential equation.

Let $\Sigma u_x = y$, then y is solely conditioned by the equation $\Delta y = u_x$, or, putting $\epsilon^{\frac{d}{dx}} - 1$ for Δ, by the linear differential equation

$$(\epsilon^{\frac{d}{dx}} - 1)\, y = u_x \quad \ldots\ldots\ldots\ldots\ldots (6).$$

Now, by the theory of linear differential equations, the complete value of y will be obtained by adding to any particular value v_x the complete value of what y would be, were u_x equal to 0. Hence

$$\Sigma u_x = v_x + C_1 \epsilon^{m_1 x} + C_2 \epsilon^{m_2 x} + \&c. \quad \ldots\ldots\ldots (7),$$

C_1, C_2, &c. being arbitrary constants, and m_1, m_2, &c. the different roots of the equation

$$\epsilon^m - 1 = 0 \quad \ldots\ldots\ldots\ldots\ldots\ldots (8).$$

Now all these roots are included in the form $m = \pm\, 2i\pi\sqrt{-1}$, i being 0 or a positive integer. When $i = 0$ we have $m = 0$, and the corresponding term in (7) reduces to a constant. But when i is a positive integer, we have in the second member of (7) a pair of terms of the form

$$C\epsilon^{2i\pi\sqrt{-1}} + C'\epsilon^{-2i\pi\sqrt{-1}},$$

which, on making $C + C' = A_i$, $(C - C')\sqrt{-1} = B_i$, is reducible to $A_i \cos 2i\pi + B_i \sin 2i\pi$. Hence, giving to i all possible integral values,

$$\Sigma u_x = v_x + C + A_1 \cos 2\pi x + A_2 \cos 4\pi x + A_3 \cos 6\pi x + \&c.$$
$$+ B_1 \sin 2\pi x + B_2 \sin 4\pi x + B_3 \sin 6\pi x + \&c.\ldots\ldots(9).$$

The portion of the right-hand member of this equation which follows v_x is the general analytical expression of a periodical constant as above defined, viz. as ever resuming the same value for values of x, whether integral or fractional, which differ by unity. It must be observed that when we have to do, as indeed usually happens, with only a particular set of values of x progressing by unity, and not with all possible sets, the periodical constant merges into an ordinary, *i.e.* into an absolute constant. Thus, if x be exclusively integral, (9) becomes

$$\Sigma u_x = v_x + C + A_1 + A_2 + A_3 + \&\mathrm{c}.$$
$$= v_x + c,$$

c being an absolute constant.

It is usual to express periodical constants of equations of differences in the form $\phi (\cos 2\pi x, \sin 2\pi x)$. But this notation is not only inaccurate, but very likely to mislead. It seems better either to employ C, leaving the interpretation to the general knowledge of the student, or to adopt the correct form

$$C + \Sigma_i (A_i \cos 2i\pi x + B_i \sin 2i\pi x) \quad \ldots\ldots\ldots(10).$$

We shall usually do the former.

Integrable Forms.

2. Of integrable forms reducible under any general law, the following are the most important.

1st Form. Factorial expressions of the form $x (x - 1) \ldots (x - m + 1)$.

Adopting the notation of Chap. II. Art. 2, we have

$$\Delta x^{(m+1)} = (m + 1) x^{(m)};$$

therefore

$$\Sigma x^{(m)} = \frac{x^{(m+1)}}{m + 1} + C,$$

or $\quad \Sigma x (x - 1) \ldots (x - m + 1) = \dfrac{x (x - 1) \ldots (x - m)}{m + 1} + C \ldots.(1).$

Thus also, if $u_x = ax + b$, we have

$$\Sigma u_x u_{x-1} \ldots u_{x-m+1} = \frac{u_x u_{x-1} \ldots u_{x-m}}{(m + 1) a} + C \ldots\ldots\ldots\ldots(2).$$

2nd Form. Rational and integral functions of x.

For, by Chap. II. Art. 5, any such function is reducible to a series of factorials of the preceding form, each of which may be integrated separately.

Ex.   To determine Σx^m.

By Chap. II. Art 5,

$$x^m = \Delta 0^m x + \frac{\Delta^2 0^m}{1.2} x^{(2)} + \frac{\Delta^3 0^m}{1.2.3} x^{(3)} \ldots + x^{(m)}.$$

Therefore

$$\Sigma x^m = C + \frac{\Delta 0^m x^{(2)}}{1.2} + \frac{\Delta^2 0^m x^{(3)}}{1.2.3} + \frac{\Delta^3 0^m x^{(4)}}{1.2.3.4} + \&c. \ldots\ldots(3).$$

In the same way we find for Σu_x the general theorem

$$\Sigma u_x = C + u_0 x + \Delta u_0 \frac{x^{(2)}}{1.2} + \Delta^2 u_0 \frac{x^{(3)}}{1.2.3} + \&c.$$

which terminates when u_x is rational and integral.

There exists also another theorem which accomplishes the same end.   We have

$$u_{x-n} = (1+\Delta)^{-n} u_x$$

$$= u_x - n\Delta u_x + \frac{n(n-1)}{1.2}\Delta^2 u_x - \&c.$$

Let $n = x$, then

$$u_0 = u_x - x\Delta u_x + \frac{x(x-1)}{1.2}\Delta^2 u_x - \&c.$$

Therefore

$$u_x = u_0 + x\Delta u_x - \frac{x(x-1)}{1.2}\Delta^2 u_x + \&c.$$

Now this being true independently of the form of u_x, we are permitted to change u_x into Σu_x. If we do this, and represent by C the value of Σu_0, we have

$$\Sigma u_x = C + x u_x - \frac{x(x-1)}{1.2}\Delta u_x + \frac{x(x-1)(x-2)}{1.2.3}\Delta^2 u_x - \&c.,$$

the theorem in question.

B. F. D.   4

It is obvious that a rational and integral function of x may also be integrated by assuming for its integral a similar function of a degree higher by unity but with arbitrary coefficients whose values are to be determined by the condition that the difference of the assumed integral shall be equal to the function given.

3rd Form. Factorial expressions of the form

$$\frac{1}{u_x u_{x+1} \cdots u_{x+m}},$$

where u_x is of the form $ax + b$.

For by Chap. II. Art. 2, we have corresponding to the above form of u_x

$$\Delta \frac{1}{u_x u_{x+1} \cdots u_{x+m-1}} = \frac{-am}{u_x u_{x+1} \cdots u_{x+m}}.$$

Hence $\qquad \Sigma \dfrac{1}{u_x u_{x+1} \cdots u_{x+m}} = -\dfrac{1}{am \left(u_x u_{x+1} \cdots u_{x+m-1} \right)}$ .....(4).

It will be observed that there must be at least two factors in the denominator of the expression to be integrated. No finite expression for $\Sigma \dfrac{1}{ax + b}$ exists.

To the above form certain more general forms are reducible.

Thus we can integrate any rational fraction of the form

$$\frac{\phi(x)}{u_x u_{x+1} \cdots u_{x+m}},$$

u_x being of the form $ax + b$, and $\phi(x)$ a rational and integral function of x of a degree lower by at least two unities than the degree of the denominator. For, expressing $\phi(x)$ in the form

$$\phi(x) = Au_x + Bu_x u_{x+1} + Cu_x u_{x+1} u_{x+2} \cdots + Eu_x u_{x+1} \cdots u_{x+m-2}.$$

$A, B\ldots$ being constants to be determined by equating coefficients, or by an obvious extension of the theorem of Chap. II. Art. 5, we find

$$\Sigma \frac{\phi(x)}{u_x u_{x+1} \cdots u_{x+m}} = A\Sigma \frac{1}{u_{x+1} u_{x+2} \cdots u_{x+m}} + B\Sigma \frac{1}{u_{x+2} u_{x+3} \cdots u_{x+m}}$$

$$\cdots + E\Sigma \frac{1}{u_{x+m-1} u_{x+m}} \cdots \cdots \cdots \cdots (5),$$

and each term can now be integrated by (4).

Again, supposing the numerator of a rational fraction to be of a degree less by at least two unities than the denominator, but intermediate factors alone to be wanting in the latter to give to it the factorial character above described, then, these factors being supplied to both numerator and denominator, the fraction may be integrated as in the last case.

Ex. Thus u_x still representing $ax + b$, we should have

$$\Sigma \frac{x}{u_x u_{x+2} u_{x+3}} = \Sigma \frac{x u_{x+1}}{u_x u_{x+1} u_{x+2} u_{x+3}},$$

with the second member of which we must proceed as before.

As all that is known of the integration of rational functions is virtually continued in the two primary theorems of (2) and (4), it is desirable to express these in the simplest form. Supposing then $u_x = ax + b$, let

$$u_x u_{x-1} \cdots u_{x-m+1} = (ax + b)^{(m)},$$

$$\frac{1}{u_x u_{x+1} \cdots u_{x+m-1}} = (ax + b)^{(-m)},$$

then

$$\Sigma (ax + b)^{(m)} = \frac{(ax + b)^{(m+1)}}{a(m + 1)} + C \cdots \cdots (6),$$

whether m be positive or negative. The analogy of this result with the theorem

$$\int (ax + b)^m \, dx = \frac{(ax + b)^{m+1}}{a(m + 1)} + C$$

is obvious.

4th Form. Functions of the form $a^x \phi(x)$ in which $\phi(x)$ is rational and integral.

4—2

Since $\Delta a^x = (a-1)\,a^x$, we have

$$\Sigma a^x = \frac{a^x}{a-1} + C.$$

To deduce $\Sigma a^x \phi(x)$ we may now employ either a method of integration by parts or a symbolical method founded upon the relations between the exponential a^x and the symbol Δ.

To integrate by parts we have,

$$\text{since } \Delta u_x v_x = u_x \Delta v_x + v_{x+1} \Delta u_x,$$

$$u_x \Delta v_x = \Delta u_x v_x - v_{x+1} \Delta u_x,$$

therefore

$$\Sigma u_x \Delta v_x = u_x v_x - \Sigma v_{x+1} \Delta u_x \dots\dots\dots\dots (7),$$

the theorem required.   Applying this we have

$$\Sigma \phi(x)\, a^x = \phi(x)\, \frac{a^x}{a-1} - \Sigma\, \frac{a^{x+1}}{a-1}\, \Delta \phi(x)$$

$$= \frac{1}{a-1}\left\{ \phi(x)\, a^x - a\Sigma a^x \Delta \phi(x) \right\}.$$

Thus the integration of $a^x \phi(x)$ is made to depend upon that of $a^x \Delta \phi(x)$; this again will by the same method depend upon that of $a^x \Delta^2 \phi(x)$, and so on.   Hence $\phi(x)$ being by hypothesis rational and integral, the process may be continued until the function under the sign Σ vanishes.   This will happen after $n+1$ operations if $\phi(x)$ be of the nth degree; and the integral will be obtained in finite terms.

But the symbolical method above referred to leads to the same result by a single operation.

By a known theorem

$$f\left(\frac{d}{dx}\right) \epsilon^{mx} \phi(x) = \epsilon^{mx} f\left(\frac{d}{dx} + m\right) \phi(x).$$

But $a^x = \epsilon^{x \log a}$.   Hence, changing in the above theorem m into $\log a$, we have

$$f\left(\frac{d}{dx}\right) a^x\, \phi(x) = a^x f\left(\frac{d}{dx} + \log a\right) \phi(x).$$

Now $\Sigma a^x \phi(x) = (\epsilon^{\frac{d}{dx}} - 1)^{-1} a^x \phi(x)$; therefore by what precedes

$$\Sigma a^x \phi(x) = a^x (\epsilon^{\frac{d}{dx} + \log a} - 1)^{-1} \phi(x)$$

$$= a^x (a \epsilon^{\frac{d}{dx}} - 1)^{-1} \phi(x)$$

$$= a^x \{a(1 + \Delta) - 1\}^{-1} \phi(x), \text{ since } \epsilon^{\frac{d}{dx}} = 1 + \Delta,$$

$$= \frac{a^x}{a-1} \left(1 + \frac{a\Delta}{a-1}\right)^{-1} \phi(x).$$

Hence developing the binomial,

$$\Sigma a^x \phi(x) = \frac{a^x}{a-1} \left\{1 - \frac{a\Delta}{a-1} + \frac{a^2\Delta^2}{(a-1)^2} - \&c.\right\} \phi(x).$$

This however is only a particular value which must be completed by the addition of an arbitrary constant. We finally get

$$\Sigma a^x \phi(x)$$

$$= \frac{a^x}{a-1} \left\{\phi(x) - \frac{a}{a-1} \Delta \phi(x) + \frac{a^2}{(a-1)^2} \Delta^2 \phi(x) - \frac{a^3}{(a-1)^3} \Delta^3 \phi(x) + \&c.\right\}..$$

$$+ C \dots\dots (8).$$

The series within the brackets stops at the nth difference of $\phi(x)$, supposing $\phi(x)$ of the nth degree.

5th Form. Functions of the forms

$$\cos(ax + b)\, \phi(x), \quad \sin(ax + b)\, \phi(x),$$

where $\phi(x)$ is rational and integral.

Beginning as before with the case in which $\phi(x) = 1$, and observing that

$$\Delta \cos(ax + b) = 2 \sin \frac{a}{2} \cos \left(ax + b + \frac{a + \pi}{2}\right),$$

$$\Delta \sin(ax + b) = 2 \sin \frac{a}{2} \sin \left(ax + b + \frac{a + \pi}{2}\right),$$

we have

$$\Sigma \cos\left(ax + b + \frac{a+\pi}{2}\right) = \frac{\cos(ax+b)}{2\sin\frac{a}{2}} + C,$$

$$\Sigma \sin\left(ax + b + \frac{a+\pi}{2}\right) = \frac{\sin(ax+b)}{2\sin\frac{a}{2}} + C.$$

Hence changing b into $b - \frac{a+\pi}{2}$, we have

$$\Sigma \cos(ax+b) = \frac{\cos\left(ax + b - \frac{a+\pi}{2}\right)}{2\sin\frac{a}{2}} + C \ \text{......} \ (9),$$

$$\Sigma \sin(ax+b) = \frac{\sin\left(ax + b - \frac{a+\pi}{2}\right)}{2\sin\frac{a}{2}} + C \text{......} (10).$$

The values of

$$\Sigma \cos(ax+b)\,\phi(x), \quad \Sigma \sin(ax+b)\,\phi(x),$$

may now be obtained, either by integration by parts, or by expressing the trigonometrical functions in their exponential forms, applying the general theorem (8), and reducing the results to a rational form.

Thus also we might integrate any function of the form $\frac{\cos}{\sin}(ax+b)\,\epsilon^{mx}\phi(x)$, or resolvable into terms of this form.

6th. *Miscellaneous Forms.* When a function proposed for integration cannot be referred to any of the preceding forms, it will be proper to divine if possible the *form* of its integral from general knowledge of the effect of the operation Δ, and to determine the constants by comparing the difference of the conjectured integral with the function proposed.

Thus since

$$\Delta a^x \phi(x) = a^x \psi(x),$$

where $\psi(x) = a\phi(x+1) - \phi(x)$, it is evident that if $\phi(x)$ be a rational fraction $\psi(x)$ will also be such. Hence if we had to integrate a function of the form $a^x\psi(x)$, $\psi(x)$ being a rational fraction, it would be proper to try first the hypothesis that the integral was of the form $a^x\phi(x)$, $\phi(x)$ being a rational fraction the constitution of which would be suggested by that of $\psi(x)$.

Ex. Thus it may be conjectured that the integral of $\dfrac{2^x(x-1)}{x(x+1)}$, if finite, will contain $\dfrac{2^x}{x}$ as a factor. And $\dfrac{2^x}{x}$ proves to be its actual value omitting the constant.

Thus also, since $\Delta \sin^{-1}\phi(x)$, $\Delta \tan^{-1}\phi(x)$, &c., are of the respective forms $\sin^{-1}\psi(x)$, $\tan^{-1}\psi(x)$, &c., $\psi(x)$ being an algebraic function when $\phi(x)$ is such, and, if $\tan^{-1}$ be employed, rational if $\phi(x)$ be so, it is usually not difficult to conjecture what must be the forms, if finite forms exist, of

$$\Sigma \sin^{-1}\psi(x), \quad \Sigma \tan^{-1}\psi(x), \quad \&c.,$$

$\psi(x)$ being still supposed algebraic.

Ex. Thus it will be found that $\tan^{-1}\dfrac{1}{p+qx+rx^2}$ is integrable in finite terms whenever the condition

$$q^2 - r^2 = 4(pr-1)$$

is satisfied. The result is

$$\Sigma \tan^{-1}\frac{1}{p+qx+rx^2} = C + \tan^{-1}\frac{2}{r-q-2rx} \quad .....(11).$$

(Herschel's *Examples*, p. 58.)

The above observations may be generalized. The operation denoted by Δ does not change or annul the *functional* characteristics of the subject to which it is applied. It does not convert transcendental into algebraic functions, or one species of transcendental functions into another. And thus, in the inverse procedure of integration, the limits of conjecture are narrowed. In the above respect the operation Δ is unlike that of differentiation, which involves essentially a procedure to the limit, and in the limit new forms arise.

Ex. Required $\Sigma \dfrac{1}{\cos ax \,.\, \cos a\,(x+1)}$.

The integral must be trigonometrical, and its factorial form shews that it must involve cos ax as a denominator. The proper form is $\dfrac{\tan ax}{\sin a} + C.$

Summation of Series.

3.   The application of the Calculus of Finite Differences to the summation of series has been already referred to in Art. 1. From what is there said it appears that to determine the sum of any portion of a series the general expression of whose terms is known, we must integrate the term which *follows* that portion, and determine the arbitrary constant by observing with what term the portion begins.

Ex.   Thus, to sum the series $1^2 + 2^2 + 3^2 \ldots + x^2$ we must first integrate $(x+1)^2$.

Now $\Sigma\,(x+1)^2 = \Sigma\,\{1 + 3x + x\,(x-1)\}$

$$= x + \frac{3x\,(x-1)}{2} + \frac{x\,(x-1)\,(x-2)}{3} + C.$$

To determine C we observe that when $x = 1$ the series is reduced to its first term 1. Hence $1 = 1 + C$, and therefore $C = 0$.

It is often more convenient to integrate the last term of that portion whose sum is required, thus obtaining the sum of the preceding terms, and then to add the last term to the result.

Thus in the above example we should have

$$1^2 + 2^2 + 3^2 \ldots + x^2 = \Sigma x^2 + x^2$$

$$= \frac{x\,(x-1)}{2} + \frac{x\,(x-1)\,(x-2)}{3} + x^2$$

on determining the constant.

It must always be carefully noted what is really the independent variable.   Thus if the series be

$$x^a + x^{a+1} + x^{a+2} \ldots + x^n,$$

the true variable is n.   Accordingly, Σ_n denoting integration with respect to n, we have

$$x^a + x^{a+1} \ldots + x^n = \Sigma_n x^{n+1} = x\Sigma_n x^n$$

$$= \frac{x^{n+1}}{x-1} + C.$$

To determine C we observe that when $n = a$ the series reduces its first term.   Hence

$$x^a = \frac{x^{a+1}}{x-1} + C;$$

$$\therefore C = x^a - \frac{x^{a+1}}{x-1} = \frac{-x^a}{x-1},$$

and we have

$$x^a + x^{a+1} \dots + x^n = \frac{x^{n+1} - x^a}{x-1},$$

the known expression.

Ex.   To sum the series $1^2 - 2^2 + 3^2 - 4^2 \dots \pm x^2$.

Here $\Sigma u_{x+1} = \Sigma (-1)^x (x+1)^2$.

Reducing by (8) Art. 2, we have finally

$$\Sigma u_{x+1} = (-1)^{x+1} \frac{x (x+1)}{2}.$$

Ex.   To sum the series $\dfrac{1}{1.3} + \dfrac{1}{2.4} + \dfrac{1}{3.5} \dots + \dfrac{1}{x (x+2)}$.

Here $\Sigma u_{x+1} = \Sigma \dfrac{1}{(x+1)(x+3)}$

$$= \Sigma \frac{x+2}{(x+1)(x+2)(x+3)}$$

$$= \Sigma \left\{ \frac{1}{(x+1)(x+2)(x+3)} + \frac{1}{(x+2)(x+3)} \right\}$$

$$= C - \frac{1}{2(x+1)(x+2)} - \frac{1}{x+2}.$$

When $x = 1$ the sum of the series is $\dfrac{1}{3}$, therefore $C = \dfrac{3}{4}$.
And on reducing

$$\Sigma u_{x+1} = \frac{3x^2 + 5x}{4(x+1)(x+2)}.$$

4.   When a series proceeds by powers of some quantity x, it frequently happens that its finite value can be obtained for some particular value of x, but not for any other value. The following is an illustration.

Ex.   To sum, when possible, the series

$$\frac{1^2 . x}{2.3} + \frac{2^2 . x^2}{3.4} + \frac{3^2 . x^3}{4.5} + \&c. \text{ to } n \text{ terms} \dots\dots\dots (a).$$

The nth term, represented by u_n, being $\dfrac{n^2 . x^n}{(n+1)(n+2)}$, we have

$$\Sigma u_{n+1} = \frac{n^2 x^n}{(n+1)(n+2)} + \Sigma \frac{n^2 x^n}{(n+1)(n+2)} \dots\dots (b).$$

Now remembering that the summation has reference to n, assume

$$\Sigma \frac{n^2 x^n}{(n+1)(n+2)} = \frac{an+b}{n+1} x^n.$$

Then, taking the difference, we have

$$\frac{x^n n^2}{(n+1)(n+2)} = x^n \left\{ x \frac{a(n+1)+b}{n+2} - \frac{an+b}{n+1} \right\}$$

$$= x^n \frac{a(x-1)n^2 + (2a+b)(x-1)n + (a+b)x - 2b}{(n+1)(n+2)}.$$

That these expressions may agree we must have

$$a(x-1) = 1, \quad (2a+b)(x-1) = 0, \quad (a+b)x - 2b = 0.$$

Whence we find

$$x = 4, \quad a = \frac{1}{3}, \quad b = -\frac{2}{3}.$$

The proposed series is therefore integrable if $x = 4$, and we have

$$\Sigma \frac{4^n . n^2}{(n+1)(n+2)} = \frac{1}{3} \frac{n-2}{n+1} . 4^n + C.$$

Substituting in (b), determining the constant, and reducing, there results

$$\frac{1^2 . 4}{2 . 3} + \frac{2^2 . 4^2}{3 . 4} \dots + \frac{n^2 4^n}{(n + 1)(n + 2)} = \frac{4^{n+1}}{3} \frac{n - 1}{n + 2} + \frac{2}{3} \dots (c).$$

Connexion of methods.

5. The series discussed in the preceding article admit also of the method of treatment developed in the treatise on Differential Equations (p. 435), and it is very interesting to compare the modes in which the same conditions of finite algebraic expression present themselves in solutions so totally distinct in form. The last example will serve as an illustration.

If we make $x = \epsilon^\theta$, the series (a) becomes

$$\frac{1^2 . \epsilon^\theta}{2 . 3} + \frac{2^2 . \epsilon^{2\theta}}{3 . 4} \dots + \frac{n^2 . \epsilon^{n\theta}}{(n + 1)(n + 2)}$$

$$= \frac{D^2}{(D + 1)(D + 2)}(\epsilon^\theta + \epsilon^{2\theta} \dots + \epsilon^{n\theta}), \text{ where } D = \frac{d}{d\theta},$$

$$= \frac{D^2}{(D + 1)(D + 2)} \frac{\epsilon^{(n+1)\theta} - \epsilon^\theta}{\epsilon^\theta - 1}$$

$$= \{1 + (D + 1)^{-1} - 4(D + 2)^{-1}\} \frac{\epsilon^{(n+1)\theta} - \epsilon^\theta}{\epsilon^\theta - 1}.$$

Effecting the integrations by means of the theorem

$$(D + a)^{-1} \phi(\theta) = \epsilon^{-a\theta} D^{-1} . \epsilon^{a\theta} \phi(\theta),$$

and then restoring x, we have for the sum of the series the expression

$$x \frac{x^n - 1}{x - 1} + \frac{1}{x} \int_0^x \frac{x^{n+1} - x}{x - 1} dx - \frac{4}{x^2} \int_0^4 \frac{x^{n+2} - x^2}{x - 1} dx.$$

Now let $x = 4$. Then substituting 4 for x in the terms without the sign of integration and in the upper limit, we have

$$4 \frac{4^n - 1}{3} + \frac{1}{4} \int_0^4 \frac{x^{n+1} - x}{x - 1} dx - \frac{1}{4} \int_0^2 \frac{x^{n+2} - x^2}{x - 1} dx$$

$$= 4\,\frac{4^n - 1}{3} + \frac{1}{4}\int_0^4 \frac{x^{n+1} - x - x^{n+2} + x^2}{x - 1}\,dx$$

$$= 4\,\frac{4^n - 1}{3} + \frac{1}{4}\int_0^4 \left(\frac{x^2 - x}{x - 1} - \frac{x^{n+2} - x^{n+1}}{x - 1}\right) dx$$

$$= 4\,\frac{4^n - 1}{3} + \frac{1}{4}\int_0^4 (x - x^{n+1})\,dx.$$

The integration can now be effected, and we have as the result

$$4\,\frac{4^n - 1}{3} + \frac{1}{4}\left(8 - \frac{4^{n+2}}{n + 2}\right)$$

$$= \frac{4^{n+1}}{3}\frac{n - 1}{n + 2} + \frac{2}{3},$$

as before.

But what is the ground of connexion between the two *methods?*

It consists in the law of reciprocity established by the theorem of Chap. II. Art. 11, viz.

$$\phi\left(\frac{d}{dt}\right)\psi\,(t) = \psi\left(\frac{d}{dt}\right)\phi\,(t),$$

$\phi\,(t)$ and $\psi\,(t)$ being functions developable by Maclaurin's theorem, and t being made equal to 0 after the implied operations are performed.

To establish this let the proposed series be

$$\phi\,(a) + \phi\,(a + 1) + \phi\,(a + 2) \ldots + \phi\,(n).$$

Its corrected value found by the method of this chapter is

$$\Sigma\phi\,(n + 1) - \Sigma\phi\,(a),$$

the summation having reference to n in the first and to a in the second term.

Its value found by the other method is

$$\phi\left(\frac{d}{dt}\right)(\epsilon^{at} + \epsilon^{(a+1)t}\ldots + \epsilon^{nt}), \text{ or } \phi\left(\frac{d}{dt}\right)\frac{\epsilon^{(n+1)t} - \epsilon^{at}}{\epsilon^t - 1},$$

where, after the implied operation, $\epsilon^t = 1$ and therefore $t = 0$.

Now by the law of reciprocity, as $t = 0$,

$$\phi\left(\frac{d}{dt}\right)\frac{\epsilon^{(n+1)t} - \epsilon^{at}}{\epsilon^t - 1} = \frac{\epsilon^{(n+1)\frac{d}{dt}} - \epsilon^{a\frac{d}{dt}}}{\epsilon^{\frac{d}{dt}} - 1}\phi(t)$$

$$= (\epsilon^{\frac{d}{dt}} - 1)^{-1}\{\epsilon^{(n+1)\frac{d}{dt}}\phi(t) - \epsilon^{a\frac{d}{dt}}\phi(t)\}$$

$$= (\epsilon^{\frac{d}{dt}} - 1)^{-1}\{\phi(t + n + 1) - \phi(t + a)\}$$

$$= (\epsilon^{\frac{d}{dn}} - 1)^{-1}\phi(t + n + 1) - (\epsilon^{\frac{d}{da}} - 1)^{-1}\phi(t+a),$$

since t is connected with n in the one term and with a in the other by the sign of addition. And now, making $t = 0$, the expression reduces to

$$\Sigma\phi(n + 1) - \Sigma\phi(a),$$

which agrees with the previous expression.

Conditions of extension of direct to inverse forms.

6. From the symbolical expression of Σ in the forms $(\epsilon^{\frac{d}{dx}} - 1)^{-1}$, and more generally of Σ^n in the form $(\epsilon^{\frac{d}{dx}} - 1)^{-n}$, flow certain theorems which may be regarded as extensions of some of the results of Chap. II. To comprehend the true nature of these extensions the peculiar *interrogative* character of the expression $(\epsilon^{\frac{d}{dx}} - 1)^{-n}u_x$ must be borne in mind. Any legitimate transformation of this expression by the development of the symbolical factor must be considered, in so far as it consists of direct forms, to be an *answer* to the question which that expression proposes; in so far as it consists of inverse forms to be a replacing of that question by others. But the answers will not be of necessity sufficiently general, and the substituted questions if answered in a perfectly unrestricted manner may lead to results which are too general. In the one case we must introduce arbitrary constants, in the other case we must determine the connecting relations among arbitrary constants; in both cases falling back upon our prior knowledge of what the character of the true solution must be. Two examples will suffice for illustration.

Ex. 1. Let us endeavour to deduce symbolically the expression for Σu_x, given in (3), Art. 1.

Now　　$\Sigma u_x = (\epsilon^{\frac{d}{dx}} - 1)^{-1} u_x$

$$= (\epsilon^{-\frac{d}{dx}} + \epsilon^{-2\frac{d}{dx}} + \&c.) \, u_x$$

$$= u_{x-1} + u_{x-2} + u_{x-3} \dots + \&c.$$

Now this is only a particular form of Σu_x corresponding to $a = -\infty$ in (3). To deduce the general form we must add an arbitrary constant, and if to that constant we assign the value

$$- (u_{a-1} + u_{a-2} \dots + \&c.),$$

we obtain the result in question.

Ex. 2. Let it be required to develope $\Sigma u_x v_x$ in a series proceeding according to Σv_x, $\Sigma^2 v_x$, &c.

We have

$$\Sigma u_x v_x = (DD' - 1)^{-1} u_x v_x,$$

D referring to x only as entering into u_x, D' to x only as entering into v_x.

Now $D' = \Delta' + 1$, therefore

$$\Sigma u_x v_x = (D\Delta' + D - 1)^{-1} u_x v_x$$

$$= (D\Delta' + \Delta)^{-1} u_x v_x$$

$$= (D^{-1}\Delta'^{-1} - D^{-2}\Delta'^{-2}\Delta + D^{-3}\Delta'^{-3}\Delta^2 - \&c.) u_x v_x$$

$$= u_{x-1}\Sigma v_x - \Delta u_{x-2}\Sigma^2 v_x + \Delta^2 u_{x-3}\Sigma^3 v_x - \&c.,$$

the theorem sought.

In applying this theorem, we are not permitted to introduce unconnected arbitrary constants into its successive terms. If we perform on both sides the operation Δ, we shall find that the equation will be identically satisfied provided $\Delta\Sigma^n u_x$ in any term is equal to $\Sigma^{n-1} u_x$ in the preceding term, and this imposes the condition that the constants in $\Sigma^{n-1} u_x$ be retained without change in $\Sigma^n u_x$. And as, if this be done, the equation will be satisfied, it follows that however many those constants may be, they will *effectively* be reduced to one. Hence then we may infer that if we express the theorem in the form

$$\Sigma u_x v_x = C + u_{x-1}\Sigma v_x - \Delta u_{x-1}\Sigma^2 v_x + \Delta^2 u_{x-2}\Sigma^3 v_x \dots \dots (1),$$

we shall be permitted to neglect the constants of integration, provided that we always deduce $\Sigma^n v_x$ by direct integration of the value of $\Sigma^{n-1} v_x$ in the preceding term.

If u_x be rational and integral, the series will be finite, and the constant C will be the one which is due to the last integration effected.

EXERCISES.

1. Find by integration the sum of the series
$$1^2 + 2^2 + 3^2 \ldots + 10^2.$$

2. Sum the series
$$\frac{5}{2 \cdot 3 \cdot 4 \cdot 5} + \frac{8}{3 \cdot 4 \cdot 5 \cdot 6} + \frac{11}{4 \cdot 5 \cdot 6 \cdot 7} + \&c. \text{ to } x \text{ terms.}$$

3. Sum the series
$$\frac{1^2 \cdot 4}{2 \cdot 3} + \frac{2^2 \cdot 4^2}{3 \cdot 4} + \frac{3^2 \cdot 4^3}{4 \cdot 5} + \frac{4^2 \cdot 4^4}{5 \cdot 6} + \&c. \text{ to } x \text{ terms.}$$

4. Prove that
$$1^3 + 2^3 + 3^3 \ldots + x^3 = (1 + 2 \ldots + x)^2.$$

5. Sum the series
$$1 + a \cos \theta + a^2 \cos 2\theta \ldots + a^{x-1} \cos (x-1)\,\theta.$$

6. The successive orders of figurate numbers are defined by this, viz. that the xth term of any order is equal to the sum of the first x terms of the order next preceding, while the terms of the first order are each equal to unity. Hence, shew that the xth term of the nth order is
$$\frac{x (x+1) (x+2) \ldots (x+n-2)}{1 \cdot 2 \ldots (n-1)}.$$

7. It is always possible to assign such values of s, real or imaginary, (being the roots of an equation of the nth degree) that the function
$$\frac{(\alpha + \beta x + \gamma x^2 \ldots + \nu x^n)\, s^x}{u_x u_{x+1} \ldots u_{x+m-1}}$$

shall be integrable in finite terms, α, β, ... ν being any constants, and u_x of the form $ax + b$.   (Herschel's *Examples of Finite Differences*, p. 47.)

8.   Sum the series

$$\frac{1}{\sin \theta} + \frac{1}{\sin 2\theta} + \frac{1}{\sin 4\theta} + \frac{1}{\sin 8\theta} + \text{\&c. to } x \text{ terms.}$$

9.   Demonstrate the following expression (different from that of Art. 6) for $\Sigma u_x v_x$, viz.

$$\Sigma u_x v_x = u_x \Sigma v_x - \Delta u_x \Sigma^2 v_{x+1} + \Delta^2 u_x \Sigma^3 v_{x+2} - \text{\&c., } ad\ inf.$$

10.   Prove the more general theorem

$$\Sigma^n u_x v_x = u_x \Sigma^n v_x - n \Delta u_x \Sigma^{n+1} v_{x+1}$$
$$+ \frac{n\,(n-1)}{1\,.\,2}\, \Delta^2 u_x \Sigma_\bullet^{n+2} v_{x+2} - \text{\&c., } ad\ inf.$$

11.   Prove the theorem

$$\Sigma u_x = C + x u_{x-1} - \frac{x\,(x-1)}{1\,.\,2}\, \Delta u_{x-2}$$
$$+ \frac{x\,(x-1)\,(x-2)}{1\,.\,2\,.\,3}\, \Delta^2 u_{x-3} - \text{\&c., } ad\ inf.$$

12.   Expand $\Sigma \phi\,(x) \cos 2mx$ in a series proceeding according to the differences of $\phi\,(x)$.

CHAPTER V.

CONVERGENCY AND DIVERGENCY OF SERIES.

1. A SERIES is said to be convergent or divergent according as the sum of its first n terms approaches or does not approach to a finite limit when n is indefinitely increased.

This definition leads us to distinguish between the convergency of a series and the convergency of the *terms* of a series. The successive terms of the series

$$1 + \frac{1}{2} + \frac{1}{3} + \frac{1}{4} + \frac{1}{5} + \&c.$$

converge to the limit 0, but it will be shewn that the sum of n of those terms tends to become infinite with n.

On the other hand, the geometrical series

$$1 + \frac{1}{2} + \frac{1}{4} + \frac{1}{8} + \frac{1}{16} + \&c.$$

is convergent both as respects its terms and as respects the sum of its terms.

2. Two cases present themselves. 1st. That in which the terms of a series are all of the same or are ultimately all of the same sign. 2ndly. That in which they are, or ultimately become, alternately positive and negative. The latter case we propose, on account of the greater simplicity of its theory, to consider first.

PROP. I. *A series whose terms diminish in absolute value, and are, or end with becoming, alternately positive and negative, is convergent.*

Let $u_1 - u_2 + u_3 - u_4 + \&c.$ be the proposed series or its terminal portion, the part which it follows being in the latter case supposed finite. Then writing it in the successive forms

$$u_1 - u_2 + (u_3 - u_4) + (u_5 - u_6) + \&c. \ldots\ldots\ldots (1),$$

$$u_1 - (u_2 - u_3) - (u_4 - u_5) - \&c. \ldots\ldots\ldots (2),$$

and observing that $u_1 - u_2$, $u_2 - u_3$, &c. are by hypothesis positive, we see that the sum of the series is greater than $u_1 - u_2$ and less than u_1. The series is therefore convergent.

Ex. Thus the series

$$1 - \frac{1}{2} + \frac{1}{3} - \frac{1}{4} + \frac{1}{5} - \&c. \ ad \ inf.$$

tends to a limit which is less than 1 and greater than $\frac{1}{2}$.

3. The theory of the convergency and divergency of series whose terms are ultimately of one sign and at the same time converge to the limit 0, will occupy the remainder of this chapter and will be developed in the following order. 1st. A fundamental proposition due to Cauchy which makes the test of convergency to consist in a process of integration, will be established. 2ndly. Certain direct consequences of that proposition relating to particular classes of series, including the geometrical, will be deduced. 3rdly. Upon those consequences, and upon a certain extension of the algebraical theory of *degree* which has been developed in the writings of Professor De Morgan and of M. Bertrand, a system of criteria general in application will be founded. It may be added that the first and most important of the criteria in question, to which indeed the others are properly supplemental, being founded upon the known properties of geometrical series, might be proved without the aid of Cauchy's proposition; but for the sake of unity it has been thought proper to exhibit the different parts of the system in their natural relation.

Fundamental Proposition.

4. PROP. *If the function* $\phi(x)$ *be positive in sign but diminishing in value as* x *varies continuously from* a *to* ∞, *then the series*

$$\phi(a) + \phi(a+1) + \phi(a+2) + \&c. \ ad \ inf \ldots\ldots (3).$$

will be convergent or divergent according as $\int_a^\infty \phi(x)\,dx$ *is finite or infinite.*

For since $\phi(x)$ diminishes from $x = a$ to $x = a + 1$, and again from $x = a + 1$ to $x = a + 2$, &c., we have

$$\int_a^{a+1} \phi(x)\,dx < \phi(a),$$

$$\int_{a+1}^{a+2} \phi(x)\,dx < \phi(a+1),$$

and so on, *ad inf.* Adding these inequations together, we have

$$\int_a^\infty \phi(x)\,dx < \phi(a) + \phi(a+1) \ldots + \text{&c. } ad\ inf.\ldots\ldots(4).$$

Again, by the same reasoning,

$$\int_a^{a+1} \phi(x)\,dx > \phi(a+1),$$

$$\int_{a+1}^{a+2} \phi(x)\,dx > \phi(a+2),$$

and so on. Again adding, we have

$$\int_a^\infty \phi(x)\,dx > \phi(a+1) + \phi(a+2) \ldots + \text{&c.}\ldots\ldots(5).$$

Thus the integral $\int_a^\infty \phi(x)\,dx$, being intermediate in value between the two series

$$\phi(a) + \phi(a+1) + \phi(a+2) \ldots$$

$$\phi(a+1) + \phi(a+2) \ldots$$

which differ by $\phi(a)$, will differ from the former series by a quantity less than $\phi(a)$, therefore by a finite quantity. Thus the series and the integral are finite or infinite together.

Cor. *If in the inequation (5) we change a into a − 1, and compare the result with (4), it will appear that the series*

$$\phi(a) + \phi(a+1) + \phi(a+2) \ldots ad\ inf.$$

has for its inferior and superior limits

$$\int_a^\infty \phi(x)\, dx, \quad \text{and} \quad \int_{u-1}^\infty \phi(x)\, dx \dots\dots\dots\dots (6).$$

The application of the above proposition will be sufficiently explained in the two following examples relating to geometrical series, and to the other classes of series involved in the demonstration of the final system of criteria referred to in Art. 3.

Ex. 1. The geometrical series

$$1 + h + h^2 + h^3 + \&c. \ ad \ inf.$$

is convergent if $h < 1$, divergent if $h \gtreqqless 1$.

The general term is h^x, the value of x in the first term being 0, so that the test of convergency is simply whether $\int_0^\infty h^x dx$ is infinite. Now

$$\int_0^x h^x dx = \frac{h^x - 1}{\log h}.$$

If $h > 1$ this expression becomes infinite with x and the series is diverging. If $h < 1$ the expression assumes the finite value $\dfrac{-1}{\log h}$. The series is therefore converging.

If $h = 1$ the expression becomes indeterminate, but, proceeding in the usual way, assumes the limiting form xh^x which becomes infinite with x. Here then the series is diverging.

Ex. 2. The successive series

$$\left.\begin{array}{l} \dfrac{1}{a^m} + \dfrac{1}{(a+1)^m} + \dfrac{1}{(a+2)^m} + \&c. \\[2ex] \dfrac{1}{a\,(\log a)^m} + \dfrac{1}{(a+1)\{\log(a+1)\}^m} + \&c. \\[2ex] \dfrac{1}{a\log a\,(\log\log a)^m} + \dfrac{1}{(a+1)\log(a+1)\{\log\log(a+1)\}^m} + \&c. \\[2ex] \dots\dots\dots\dots\dots\dots\dots\dots\dots\dots\dots\dots\dots\dots \end{array}\right\} (7),$$

a being positive, are convergent if $m > 1$, and divergent if $m \lessgtr 1$.

The determining integrals are

$$\int_a^\infty \frac{dx}{x^m}, \quad \int_a^\infty \frac{dx}{x(\log x)^m}, \quad \int_a^\infty \frac{dx}{x\log x (\log\log x)^m}, \ldots\ldots$$

and their values, except when m is equal to 1, are

$$\frac{x^{1-m}-a^{1-m}}{1-m}, \quad \frac{(\log x)^{1-m}-(\log a)^{1-m}}{1-m}, \quad \frac{(\log\log x)^{1-m}-(\log\log a)^{1-m}}{1-m} \ldots$$

in which $x = \infty$. All these expressions are infinite if m be less than 1, and finite if m be greater than 1. If $m = 1$ the integrals assume the forms

$\log x - \log a$, $\log\log x - \log\log a$, $\log\log\log x - \log\log\log a$ &c.

and still become infinite with x. Thus the series are convergent if $m > 1$ and divergent if $m \lessgtr 1$.

5. Perhaps there is no other mode so satisfactory for establishing the convergency or divergency of a series as the direct application of Cauchy's proposition, when the integration which it involves is possible. But, as this is not always the case, the construction of a system of derived rules not involving a process of integration becomes important. To this object we now proceed.

First derived Criterion.

PROP. *The series $u_0 + u_1 + u_2 \ldots$ ad inf., all whose terms are supposed positive, is convergent or divergent according as the ratio $\dfrac{u_{x+1}}{u_x}$ tends, when x is indefinitely increased, to a limiting value less or greater than unity.*

Let h be that limiting value; and first let h be less than 1, and let k be some positive quantity so small that $h + k$ shall also be less than 1. Then as $\dfrac{u_{x+1}}{u_x}$ tends to the limit h, it is possible to give to x some value n so large, yet finite, that for that value and for all superior values of x the ratio $\frac{u_{x+1}}{u_x}$ shall lie within the limits $h+k$ and $h-k$. Hence if, beginning with the particular value of x in question, we construct the three series

$$\left.\begin{array}{l} u_n + (h+k)\,u_n + (h+k)^2\,u_n + \&c. \\ u_n + u_{n+1} \quad\quad + \quad u_{n+2} \quad + \&c. \\ u_n + (h-k)\,u_n + (h-k)^2\,u_n + \&c. \end{array}\right\} \ldots\ldots\ldots\ldots (8),$$

each term after the first in the second series will be intermediate in value between the corresponding terms in the first and third series, and therefore the second series will be intermediate in value between

$$\frac{u_n}{1-(h+k)} \text{ and } \frac{u_n}{1-(h-k)},$$

which are the finite values of the first and third series. And therefore the given series is convergent.

On the other hand, if h be greater than unity, then, giving to k some small positive value such that $h-k$ shall also exceed unity, it will be possible to give to x some value n so large, yet finite, that for that and all superior values of x, $\frac{u_{x+1}}{u_x}$ shall lie between $h+k$ and $h-k$. Here then still each term after the first in the second series will be intermediate between the corresponding terms of the first and third series. But $h+k$ and $h-k$ being both greater than unity, both the latter series are divergent (Ex. 1). Hence the second or given series is divergent also.

Ex. 3.   The series $1 + t + \frac{t^2}{1.2} + \frac{t^3}{1.2.3} + \&c.$, derived from the expansion of ϵ^t, is convergent for all values of t.

For if

$$u_x = \frac{t^x}{1.2\ldots x}, \quad u_{x+1} = \frac{t^{x+1}}{1.2\ldots(x+1)},$$

then $$\frac{u_{x+1}}{u_x} = \frac{t}{x+1},$$

and this tends to 0 as x tends to infinity.

Ex. 4. The series

$$1 + \frac{a}{b}t + \frac{a(a+1)}{b(b+1)}t^2 + \frac{a(a+1)(a+2)}{b(b+1)(b+2)}t^3 + \&c. \ldots\ldots (9),$$

is convergent or divergent according as t is less or greater than unity.

Here $$u_x = \frac{a(a+1)(a+2)\ldots(a+x-1)}{b(b+1)(b+2)\ldots(b+x-1)}t^x,$$

Therefore $$\frac{u_{x+1}}{u_x} = \frac{a+x}{b+x}t,$$

and this tends, x being indefinitely increased, to the limit t. Accordingly therefore as t is less or greater than unity, the series is convergent or divergent.

If $t = 1$ the rule fails. Nor would it be easy to apply directly Cauchy's test to this case, because of the indefinite number of factors involved in the expression of the general term of the series. We proceed, therefore, to establish the supplemental criteria referred to in Art. 3.

Supplemental Criteria.

6. Let the series under consideration be

$$u_a + u_{a+1} + u_{a+2} + u_{a+3} \ldots ad\ inf. \ldots\ldots\ldots (10),$$

the general term u_x being supposed positive and diminishing in value from $x = a$ to $x = $ infinity. The above form is adopted as before to represent the terminal, and by hypothesis positive, portion of series whose terms do not necessarily begin with being positive; since it is upon the character of the terminal portion that the convergency or divergency of the series depends.

It is evident that the series (10) will be convergent if its terms become ultimately less than the corresponding terms of a known convergent series, and that it will be divergent if its terms become ultimately greater than the corresponding terms of a known divergent series.

Compare then the above series whose general term is u_x with the first series in (7), Ex. 2, whose general term is $\dfrac{1}{x^m}$. Then a condition of convergency is

$$u_x < \frac{1}{x^m},$$

m being greater than unity, and x being indefinitely increased.

Hence we find

$$x^m < \frac{1}{u_x};$$

$$\therefore\ m \log x < \log \frac{1}{u_x};$$

$$m < \frac{\log \dfrac{1}{u_x}}{\log x},$$

and since m is greater than unity

$$\frac{\log \dfrac{1}{u_x}}{\log x} > 1.$$

On the other hand, there is divergency if

$$u_x > \frac{1}{x^m},$$

x being indefinitely increased, and m being equal to or less than 1. But this gives

$$m > \frac{\log \dfrac{1}{u_x}}{\log x},$$

and therefore

$$\frac{\log \dfrac{1}{u_x}}{\log x} < 1.$$

It appears therefore that the series *is convergent or divergent according as, x being indefinitely increased, the function* $\dfrac{log\,\dfrac{1}{u_x}}{log\,x}$ *approaches a limit greater or less than unity.*

But the limit being unity, and the above test failing, let the comparison be made with the second of the series in (7). For convergency, we then have as the limiting equation,

$$u_x < \frac{1}{x\,(\log x)^m},$$

m being greater than unity. Hence we find, by proceeding as before,

$$\frac{\log \dfrac{1}{xu_x}}{\log \log x} > 1.$$

And deducing in like manner the condition of divergency, we conclude that *the series is convergent or divergent according as, x being indefinitely increased, the function* $\dfrac{log\,\dfrac{1}{xu_x}}{log\,log\,x}$ *tends to a limit greater or less than unity.*

Should this limit be unity, we must have recourse to the third series of (7), the resulting test being that *the proposed series is convergent or divergent according as, x being indefinitely increased, the function* $\dfrac{log\,\dfrac{1}{x\,log\,xu_x}}{log\,log\,log\,x}$ *tends to a limit greater or less than unity.*

The forms of the functions involved in the succeeding tests, *ad inf.*, are now obvious. Practically, we are directed to construct the successive functions,

$$\frac{l\,\dfrac{1}{u_x}}{lx},\quad \frac{l\,\dfrac{1}{xu_x}}{llx},\quad \frac{l\,\dfrac{1}{x\,lx\,u_x}}{lllx},\quad \frac{l\,\dfrac{1}{x\,lx\,llx\,u_x}}{llllx},\quad \&\text{c.} \ldots\ldots\ldots (A),$$

and the first of these which tends, as *x* is indefinitely increased to a limit greater or less than unity, determines the series to be convergent or divergent.

The criteria may be presented in another form. For representing $\dfrac{1}{u_x}$ by $\phi(x)$, and applying to each of the functions in (A), the rule for indeterminate functions of the form $\dfrac{\infty}{\infty}$, we have

$$\frac{l\phi(x)}{lx} = \frac{\phi'(x)}{\phi(x)} \div \frac{1}{x} = \frac{x\phi'(x)}{\phi(x)},$$

$$\frac{l\dfrac{\phi(x)}{x}}{llx} = \left\{\frac{\phi'(x)}{\phi(x)} - \frac{1}{x}\right\} \div \frac{1}{x\log x}$$

$$= \log x \left\{x\frac{\phi'(x)}{\phi(x)} - 1\right\},$$

and so on. Thus the system of functions (A) is replaced by the system

$$\frac{x\phi'(x)}{\phi(x)}, \quad lx\left\{\frac{x\phi'(x)}{\phi(x)} - 1\right\},$$

$$llx\left[lx\left\{x\frac{\phi'(x)}{\phi(x)} - 1\right\} - 1\right], \&c. \ \ldots\ldots\ (B).$$

It was virtually under this form that the system of functions was originally presented by Prof. De Morgan, (*Differential Calculus*, pp. 325–7). The law of formation is as follows. If P_n represent the nth function, then

$$P_{n+1} = l^n x\,(P_n - 1)\ \ldots\ldots\ldots\ldots\ldots\ (11).$$

7. There exists yet another and equivalent system of determining functions which in particular cases possesses great advantages over the two above noted. It is obtained by substituting in Prof. De Morgan's forms $\dfrac{u_x}{u_{x+1}} - 1$ for $\dfrac{\phi'(x)}{\phi(x)}$ The lawfulness of this substitution may be established as follows.

Since $\qquad u_x = \dfrac{1}{\phi(x)}$, we have

$$\frac{u_x}{u_{x+1}} - 1 = \frac{\phi(x+1)}{\phi(x)} - 1$$

$$= \frac{\phi(x+1) - \phi(x)}{\phi(x)}.$$

Now by a known theorem on the limits of Taylor's series,

$$\phi(x+1) - \phi(x) = \phi'(x+\theta),$$

θ being some quantity between 0 and 1. Hence

$$\frac{u_x}{u_{x+1}} - 1 = \frac{\phi'(x+\theta)}{\phi(x)}$$

$$= \frac{\phi'(x)}{\phi(x)} \frac{\phi'(x+\theta)}{\phi'(x)} \dots\dots\dots\dots (12).$$

Now $\dfrac{\phi'(x+\theta)}{\phi'(x)}$ has 1 for its limiting value; for, since u_x by hypothesis converges to 0 as x is indefinitely increased, $\phi(x)$ tends to become infinite, and therefore $\dfrac{\phi(x+\theta)}{\phi(x)}$ assumes the form $\dfrac{\infty}{\infty}$; therefore

$$\frac{\phi(x+\theta)}{\phi(x)} = \frac{\phi'(x+\theta)}{\phi'(x)}.$$

But the first member has for its limits $\dfrac{\phi(x)}{\phi(x)}$ and $\dfrac{\phi(x+1)}{\phi(x)}$, i.e. 1 and $\dfrac{\phi(x+1)}{\phi(x)}$; and the latter ratio tends to unity as x is indefinitely increased, since by hypothesis $\dfrac{u_{x+1}}{u_x}$ tends to that limit. Hence $\dfrac{\phi'(x+\theta)}{\phi'(x)}$ tends to the limit 1. Thus (12) becomes

$$\frac{u_x}{u_{x+1}} - 1 = \frac{\phi'(x)}{\phi(x)}.$$

Substituting therefore in (B), we obtain the system of functions

$$x\left(\frac{u_x}{u_{x+1}} - 1\right), \quad lx\left\{x\left(\frac{u_x}{u_{x+1}} - 1\right) - 1\right\},$$

$$llx \left[lx \left\{ x \left(\frac{u_x}{u_{x+1}} - 1 \right) - 1 \right\} \right] \&\text{c.} \ldots \ldots \ldots (C),$$

the law of formation being still $P_{n+1} = l^n x \, (P_n - 1)$.

8. The results of the above inquiry may be collected into the following rule.

RULE. *Determine first the limiting value of the function* $\frac{u_{x+1}}{u_x}$. *According as this is less or greater than unity the series is convergent or divergent.*

But if that limiting value be unity, seek the limiting values of whichsoever is most convenient of the three systems of functions (A), (B), (C). *According as, in the system chosen, the first function whose limiting value is not unity, assumes a limiting value greater or less than unity, the series is convergent or divergent.*

Ex. 5. Let the given series be

$$1 + \frac{1}{2^{\frac{2}{2}}} + \frac{1}{3^{\frac{4}{3}}} + \frac{1}{4^{\frac{5}{4}}} + \&\text{c.} \ldots \ldots \ldots \ldots (13).$$

Here $u_x = \dfrac{1}{x^{\frac{x+1}{x}}}$,   therefore,

$$\frac{u_{x+1}}{u_x} = \frac{x^{\frac{x+1}{x}}}{(x+1)^{\frac{x+2}{x+1}}} = \frac{x^{1+\frac{1}{x}}}{(x+1)^{1+\frac{1}{x+1}}},$$

and x being indefinitely increased the limiting value is unity.

Now applying the first criterion of the system (A), we have

$$\frac{l \dfrac{1}{u_x}}{lx} = \frac{\dfrac{x+1}{x} lx}{lx} = \frac{x+1}{x},$$

and the limiting value is again unity. Applying the second criterion in (A), we have

$$\frac{l \dfrac{1}{x u_x}}{llx} = \frac{lx^{\frac{1}{x}}}{llx} = \frac{lx}{x llx},$$

the limiting value of which found in the usual way is 0. Hence the series is divergent.

Ex. 6.   Resuming the hypergeometrical series of Ex. 4, viz.

$$1 + \frac{a}{b}t + \frac{a(a+1)}{b(b+1)}t^2 + \frac{a(a+1)(a+2)}{b(b+1)(b+2)}t^3 + \&c\ldots (14),$$

we have, in the failing case of $t = 1$,

$$u_x = \frac{a(a+1)\ldots(a+x-1)}{b(b+1)\ldots(b+x-1)}.$$

Therefore

$$\frac{u_{x+1}}{u_x} = \frac{a+x}{b+x},$$

and applying the first criterion of (C),

$$x\left(\frac{u_x}{u_{x+1}} - 1\right) = x\left(\frac{b+x}{a+x} - 1\right)$$

$$= \frac{(b-a)x}{a+x},$$

which tends to the limit $b - a$. The series is therefore convergent or divergent according as $b - a$ is greater or less than unity.

If $b - a$ is equal to unity, we have, by the second criterion of (C),

$$lx\left\{x\left(\frac{u_x}{u_{x+1}} - 1\right) - 1\right\} = lx\left\{\frac{(b-a)x}{a+x} - 1\right\}$$

$$= \frac{-alx}{a+x},$$

since $b - a = 1$. The limiting value is 0, so that the series is still divergent.

It appears, therefore, 1st, that the series (14) is convergent or divergent according as t is less or greater than 1; 2ndly, that if $t = 1$ the series is convergent if $b - a > 1$, divergent if $b - a \lessgtr 1$.

9.   The extension of the theory of *degree* referred to in Art. 3, is involved in the demonstration of the above criteria.

When two functions of x are, in the ordinary sense of the term, of the same degree, i. e. when they respectively involve the same highest powers of x, they tend, x being indefinitely increased, to a ratio which is finite yet not equal to 0; viz. to the ratio of the respective coefficients of that highest power. Now let the converse of this proposition be assumed as the *definition* of equality of degree, i.e. let *any* two functions of x be said to be of the same degree when the ratio between them tends, x being indefinitely increased, to a finite limit which is not equal to 0. Then are the several functions

$$x\,(lx)^m, \qquad xlx\,(llx)^m, \qquad \&c.,$$

with which $\dfrac{1}{u_x}$ or $\phi\,(x)$ is successively compared in the demonstrations of the successive criteria, so many interpositions of degree between x and x^{1+a}, however small a may be. For x being indefinitely increased, we have

$$\lim \frac{x\,(lx)^m}{x} = \infty\,, \quad \lim \frac{x\,(lx)^m}{x^{1+a}} = 0,$$

$$\lim \frac{xlx\,(llx)^m}{xlx} = \infty\,, \quad \lim \frac{xlx\,(llx)^m}{x^{1+a}} = 0,$$

so that, according to the definition, $x\,(lx)^m$ is intermediate in degree between x and x^{1+a}, $xlx\,(llx)^m$ between xlx and $x\,(lx)^{1+a}$, &c. And thus each failing case, arising from the supposition of $m = 1$, is met by the introduction of a new function.

It may be noted in conclusion that the first criterion of the system (A) was originally demonstrated by Cauchy, and the first of the system (C) by Raabe (*Crelle*, Vol. IX.). Bertrand*, to whom the comparison of the three systems is due, has demonstrated that if one of the criteria should fail from the absence of a *definite* limit, the succeeding criteria will also fail in the same way. The possibility of their continued failure through the continued reproduction of the definite limit 1, is a question which has indeed been noticed but has scarcely been discussed.

---

* Liouville's *Journal*, Tom. VII. p. 35.

EXERCISES.

1. Find, by an application of the fundamental proposition, Art. 4, two limits of the value of the series

$$\frac{1}{a^2} + \frac{1}{a^2+1} + \frac{1}{a^2+4} + \frac{1}{a^2+9} + \&c. \ ad \ inf.$$

In particular shew that if $a = 1$, the numerical value of the series will lie between the limits $\dfrac{\pi}{2}$ and $\dfrac{\pi}{4}$.

2. The series $x - \dfrac{x^2}{2} + \dfrac{x^3}{3} - \dfrac{x^4}{4} + \&c.$ is convergent if x is equal to or less than unity, divergent if x is greater than unity.

3. Prove, from the fundamental proposition, Art. 4, that the two series

$$\left. \begin{array}{l} \phi(1) + \phi(2) + \phi(3) + \&c. \ ad \ inf. \\ \phi(1) + m\phi(m) + m^2\phi(m^2) + \&c. \ ad \ inf. \end{array} \right\}, \ m \text{ being positive}$$

are convergent or divergent together.

4. The hypergeometrical series

$$1 + \frac{ab}{cd}x + \frac{a(a+1)b(b+1)}{c(c+1)d(d+1)}x^2 + \&c.$$

is convergent if $x < 1$, divergent if $x > 1$.

If $x = 1$ it is convergent if $c + d - a - b > 1$, divergent if $c + d - a - b \lesseqgtr 1$.

5. The series $1 + 2^a + \dfrac{3^a}{2^{a+b}} + \dfrac{4^a}{3^{a+b}} + \&c.$ is convergent if $b > 1$, divergent if $b \lesseqgtr 1$.

CHAPTER VI.

THE APPROXIMATE SUMMATION OF SERIES.

1. IT has been seen that the finite summation of series depends upon our ability to express in finite algebraical terms, the result of the operation Σ performed upon the general term of the series. When such finite expression is beyond our powers, theorems of approximation must be employed. And the constitution of the symbol Σ as expressed by the equation

$$\Sigma = (\epsilon^{\frac{d}{dx}} - 1)^{-1} \dots (1),$$

renders the deduction and the application of such theorems easy.

Speaking generally these theorems are dependent upon the development of the symbol Σ in ascending powers of $\frac{d}{dx}$, or, under particular circumstances, in ascending powers of Δ. Hence two classes of theorems arise, viz. 1st, those which express the sum of a series whose general term is given, by a new and rapidly convergent series proceeding according to the differential coefficients of the general term, or to the differential coefficients of some important factor of the general term; 2ndly, those which differ from the above only in that *differences* take the place of differential coefficients. The former class of theorems is the more important, but examples of both and illustrations of their use will be given.

2. PROP. *To develope Σu_x in a series proceeding by the differential coefficients of u_x.*

Since $\Sigma u_x = (\epsilon^{\frac{d}{dx}} - 1)^{-1} u_x$, we must expand $(\epsilon^{\frac{d}{dx}} - 1)^{-1}$ in ascending powers of $\frac{d}{dx}$, and the *form* of the expansion will be determined by that of the function $(\epsilon^t - 1)^{-1}$. For simplicity we will first deduce a few terms of the expansion and afterwards determine its general law. Now

$$(\epsilon^t - 1)^{-1} = \cfrac{1}{t + \dfrac{t^2}{2} + \dfrac{t^3}{2 \cdot 3} + \dfrac{t^4}{2 \cdot 3 \cdot 4} + \&c.}$$

$$= \frac{1}{t} - \frac{1}{2} + \frac{t}{12} - \frac{t^3}{720} + \&c. \ldots \ldots \ldots (1),$$

on actual division. Hence replacing t by $\dfrac{d}{dx}$ and restoring the subject u_x, we have

$$\Sigma u_x = \left(\frac{d}{dx}\right)^{-1} u_x - \frac{1}{2} u_x + \frac{1}{12} \frac{du_x}{dx} - \frac{1}{720} \frac{d^3 u_x}{dx^3} + \&c.$$

$$= C + \int u_x dx - \frac{1}{2} u_x + \frac{1}{12} \frac{du_x}{dx} - \frac{1}{720} \frac{d^3 u_x}{dx^3} + \&c. \ldots \ldots (2).$$

For most applications this would suffice, but we shall proceed to determine the law of the series.

The development of the function $(\epsilon^t - 1)^{-1}$ cannot be directly obtained by Maclaurin's theorem, since, as appears from (1), it contains a negative index; but it may be obtained by expanding $\dfrac{t}{\epsilon^t - 1}$ by Maclaurin's theorem and dividing the result by t.

Referring to (1) we see that the development of $\dfrac{t}{\epsilon^t - 1}$ will have $-\frac{1}{2}t$ for its second term. It will now be shewn that this is the only term of the expansion which involves an odd power of t. Let

$$\frac{t}{\epsilon^t - 1} = -\frac{1}{2}t + R,$$

R being the sum of all the other terms of the expansion. Then

$$R = \frac{t}{\epsilon^t - 1} + \tfrac{1}{2}t = \tfrac{1}{2}t\,\frac{\epsilon^t + 1}{\epsilon^t - 1}$$

$$= \frac{t}{2}\,\frac{\epsilon^{\frac{t}{2}} + \epsilon^{-\frac{t}{2}}}{\epsilon^{\frac{t}{2}} - \epsilon^{-\frac{t}{2}}},$$

and as this does not change on changing t into $-t$, the terms represented by R contain only even powers of t.

Now if for the moment we represent $\epsilon^t - 1$ by θ, we have

$$\frac{t}{\epsilon^t - 1} = \frac{\log(1 + \theta)}{\theta} = 1 - \frac{\theta}{2} + \frac{\theta^2}{3} - \frac{\theta^3}{4} + \&c.$$

$$= 1 - \frac{\epsilon^t - 1}{2} + \frac{(\epsilon^t - 1)^2}{3} - \frac{(\epsilon^t - 1)^3}{4} + \&c. \ldots\ldots (3).$$

But by the secondary form of Maclaurin's theorem, Chap. II. Art. 11,

$$(\epsilon^t - 1)^n = \frac{\Delta^n 0^n}{n\rceil}\,t^n + \frac{\Delta^n 0^{n+1}}{n+1\rceil}\,t^{n+1} + \frac{\Delta^n 0^{n+2}}{n+2\rceil}\,t^{n+2} + \&c.,$$

in which the coefficient of t^m is $\dfrac{\Delta^n 0^m}{m\rceil}$. Hence the coefficient of t^m in (3) is

$$\frac{1}{m\rceil}\left\{0^m - \tfrac{1}{2}\Delta 0^m + \tfrac{1}{3}\Delta^2 0^m \ldots + \frac{(-1)^m}{m+1}\,\Delta^m 0^m\right\} \ldots\ldots (4),$$

since $\Delta^n 0^m$ vanishes when n is greater than m. It is to be noted that when $m = 0$ we have $0^m = 1$ and $\overline{m\rceil} = 1$.

The expression (4) determines in succession *all* the coefficients of the development of $t\,(\epsilon^t - 1)^{-1}$ in ascending powers of t. It must therefore, and it will, vanish when m receives any odd value greater than 1.

From these results we may conclude that the development of $(\epsilon^t - 1)^{-1}$ will assume the form

$$(\epsilon^t - 1)^{-1} = \frac{A_{-1}}{t} + A_0 + A_1 t + A_3 t^3 + A_5 t^5 + \&c. \ldots\ldots (5),$$

where in general A_{m-1} is expressed by (4).

It is however customary to express this development in the somewhat more arbitrary form

$$(\epsilon^t - 1)^{-1} = \frac{1}{t} - \frac{1}{2} + \frac{B_1}{2!}\, t - \frac{B_3}{4!}\, t^3 + \frac{B_5}{6!}\, t^5 - \&c. \ldots\ldots (6).$$

The quantities B_1, B_3, &c. are called Bernoulli's numbers, and their general expression will evidently be

$$B_{2r-1} = (-1)^{r+1} \left\{ -\tfrac{1}{2}\Delta 0^{2r} + \tfrac{1}{3}\Delta^2 0^{2r} \ldots + \frac{\Delta^{2r} 0^{2r}}{2r+1} \right\} \ldots\ldots (7).$$

Hence we find

$$\Sigma u_x = C + \int u_x dx - \tfrac{1}{2} u_x + \frac{B_1}{2!}\frac{du_x}{dx} - \frac{B_3}{4!}\frac{d^3 u_x}{dx^3} + \&c. \ldots\ldots (8).$$

Or, actually calculating a few of the coefficients by means of the table of the differences of 0 given in Chap. II.,

$$\Sigma u_x = C + \int u_x dx - \frac{1}{2} u_x + \frac{1}{12}\frac{du_x}{dx} - \frac{1}{720}\frac{d^3 u_x}{dx^3}$$

$$+ \frac{1}{30240}\frac{d^5 u_x}{dx^5} \ldots\ldots (9).$$

3.   Before proceeding to apply the above theorem a few observations are necessary.

Attention has been directed (*Differential Equations*, p. 376) to the *interrogative* character of inverse forms such as

$$(\epsilon^{\frac{d}{dx}} - 1)^{-1} u_x.$$

The object of a theorem of transformation like the above is, strictly speaking, to determine a function of x such that if we perform upon it the corresponding *direct* operation (in the above case this is $\epsilon^{\frac{d}{dx}} - 1$) the result will be u_x. To the inquiry what that function is, a legitimate transformation will necessarily give a correct but not necessarily the most general answer. Thus C in the second member of (8) is, from the mode of its introduction, the constant of ordinary integration; but for the most general expression of Σu_x C ought to be a periodical quantity, subject only to the condition of resuming the same value for values of x differing by unity. In the applications to which we shall proceed the values of x involved will be-integral, so that it will suffice to regard C as a simple constant. Still it is important that the true relation of the two members of the equation (8) should be understood.

The following table contains the values of the first ten of Bernoulli's numbers calculated from (7),

$$B_1 = \frac{1}{6}, \ B_3 = \frac{1}{30}, \ B_5 = \frac{1}{42}, \ B_7 = \frac{1}{30}, \ B_9 = \frac{5}{66},$$

$$B_{11} = \frac{691}{2730}, \ B_{13} = \frac{7}{6}, \ B_{15} = \frac{3617}{510},$$

$$B_{17} = \frac{43867}{798}, \ B_{19} = \frac{1222277}{2310} \ \dots\dots\dots (10).$$

It will be noted that they are ultimately divergent. It can seldom however be necessary to carry the series for Σu_x further than is done in (9), and it will be shewn that the employment of its convergent portion is sufficient.

Applications.

4.   The general expression for Σu_x in (9), Art. 2, gives us at once the integral of any rational and entire function of x.

Ex. 1.   Thus making $u_x = x^4$, we have

$$\Sigma x^4 = C + \int x^4 dx - \frac{1}{2} x^4 + \frac{1}{12} \frac{d(x^4)}{dx} - \frac{1}{720} \frac{d^3(x^4)}{dx^3}$$

$$= C + \frac{x^5}{5} - \frac{x^4}{2} + \frac{x^3}{3} - \frac{x}{30}.$$

But the theorem is of chief importance when finite summation is impossible.

Ex. 2.   Thus making $u_x = \frac{1}{x^3}$, we have

$$\Sigma \frac{1}{x^2} = C - \frac{1}{x} - \frac{1}{2x^2} + \frac{1}{12}\left(\frac{-2}{x^3}\right) - \frac{1}{720}\left(-\frac{2.3.4}{x^5}\right) - \&c.$$

$$= C - \frac{1}{x} - \frac{1}{2x^2} - \frac{1}{6x^3} + \frac{1}{30x^5} - \&c.$$

The value of C must be determined by the particular conditions of the problem. Thus suppose it required to determine an approximate value of the series

$$\frac{1}{1^2} + \frac{1}{2^2} + \frac{1}{3^2} \cdots + \frac{1}{(x-1)^2}.$$

Now by what precedes,

$$\frac{1}{1^2} + \frac{1}{2^2} + \frac{1}{3^2} \cdots + \frac{1}{(x-1)^2} = C - \frac{1}{x} - \frac{1}{2x^2} - \frac{1}{6x^3} + \frac{1}{30x^5}, \&c.$$

Let $x = \infty$, then the first member is equal to $\frac{\pi^2}{6}$ by a known theorem, while the second member reduces to C. Hence

$$\frac{1}{1^2} + \frac{1}{2^2} \cdots + \frac{1}{(x-1)^2} = \frac{\pi^2}{6} - \frac{1}{x} - \frac{1}{2x^2} - \frac{1}{6x^3} + \frac{1}{30x^5}, \&c.$$

and if x be large a few terms of the series in the second member will suffice.

5. When the sum of the series *ad inf.* is unknown, or is known to be infinite, we may approximately determine C by giving to x some value which will enable us to compare the expression for Σu_x in which the constant is involved with the actual value of Σu_x obtained from the given series by addition of its terms.

Ex. 3.   Let the given series be $1 + \frac{1}{2} + \frac{1}{3} \cdots + \frac{1}{x}$.

Representing this series by u, we have

$$u = \frac{1}{x} + \Sigma \frac{1}{x}$$

$$= \frac{1}{x} + C + \log x - \frac{1}{2x} - \frac{1}{12x^2} + \frac{1}{120x^4} + \&c.$$

$$= C + \log x + \frac{1}{2x} - \frac{1}{12x^2} + \frac{1}{120x^4}.$$

To determine C, assume $x = 10$, then

$$1 + \frac{1}{2} + \frac{1}{3} \ldots + \frac{1}{10} = C + \log_e 10 + \frac{1}{20} - \frac{1}{1200} + \frac{1}{1200000} + \&\text{c}.$$

Hence, writing for $\log_e 10$ its value $2 \cdot 30258$, we have approximately $C = \cdot 577215$. Therefore

$$u = \cdot 577215 + \log_e x + \frac{1}{2x} - \frac{1}{12x^2} + \frac{1}{120x^4} - \&\text{c}.$$

Ex.   Required an approximate value for $1 \cdot 2 \cdot 3 \ldots x$.

If $u = 1 \cdot 2 \cdot 3 \ldots x$, we have

$$\log u = \log 1 + \log 2 + \log 3 \ldots + \log x$$
$$= \log x + \Sigma \log x.$$

But $\Sigma \log x = C + \int \log x \, dx - \frac{1}{2} \log x$

$$+ \frac{B_1}{1.2} \frac{d \log x}{dx} - \frac{B_3}{1.2.3.4} \frac{d^3 \log x}{dx^3} + \&\text{c}.$$

$$= C + \left(x - \frac{1}{2}\right) \log x - x + \frac{B_1}{1.2x} - \frac{B_3}{3.4x^3} + \frac{B_5}{5.6x^5} - \&\text{c}.;$$

$$\therefore \log u = C + \left(x + \frac{1}{2}\right) \log x - x + \frac{1}{12x} - \&\text{c}. \ldots \ldots (11).$$

To determine C, suppose x very large and tending to become infinite, then

$$\log (1 \cdot 2 \cdot 3 \ldots x) = C + \left(x + \frac{1}{2}\right) \log x - x,$$

whence

$$1 \cdot 2 \cdot 3 \ldots x = \epsilon^{C-x} \times x^{x+\frac{1}{2}} \ldots \ldots \ldots \ldots \ldots (12),$$

$$1 \cdot 2 \cdot 3 \ldots 2x = \epsilon^{C-2x} \times (2x)^{2x+\frac{1}{2}} \ldots \ldots \ldots \ldots (13).$$

But, multiplying (12) by 2^x,

$$2 \cdot 4 \cdot 6 \ldots 2x \doteq 2^x \, \epsilon^{C-x} \times x^{x+\frac{1}{2}} \ldots \ldots \ldots \ldots (14).$$

Therefore, dividing (13) by (14),

$$3 . 5 . 7 \ldots (2x - 1) = 2^{x+\frac{1}{2}} \, \epsilon^{-x} x^{x},$$

whence
$$\frac{2 . 4 . 6 \ldots 2x}{3 . 5 . 7 \ldots (2x - 1)} = \frac{\epsilon^{c} x^{\frac{1}{2}}}{2^{\frac{1}{2}}} \, .$$

But by Wallis's theorem, x being infinite,

$$\frac{2 . 4 . 6 \ldots (2x - 2) \sqrt{(2x)}}{3 . 5 . 7 \ldots (2x - 1)} = \sqrt{\left(\frac{\pi}{2}\right)},$$

whence by division

$$\sqrt{(2x)} = \frac{\epsilon^{c} \sqrt{x}}{\sqrt{\pi}} \, ;$$

$$\therefore \; C = \log \sqrt{(2\pi)}.$$

And now, substituting this value in (11) and determining u, we find

$$u = \sqrt{(2\pi)} \times x^{x+\frac{1}{2}} \times \epsilon^{-x + \frac{1}{12x} - \frac{1}{360x^3} + \&c.}$$

$$= \sqrt{(2\pi x)} . \, x^{x} . \, \epsilon^{-x + \frac{1}{12x} - \frac{1}{360x^3} + \&c.} \ldots\ldots\ldots\ldots (15),$$

If we develope the factor $\epsilon^{\frac{1}{12x} - \frac{1}{360x^3} + \&c.}$ in descending powers of x, we find

$$1 . 2 . 3 \ldots x = \sqrt{(2\pi x)} . x^{x} \epsilon^{-x} \left(1 + \frac{1}{12x} + \frac{1}{288x^2} - \frac{139}{51840x^3} + \&c. \right)$$

$$\ldots\ldots\ldots (16).$$

Hence for very large values of x we may assume

$$1 . 2 . 3 \ldots x = \sqrt{(2\pi x)} \left(\frac{x}{\epsilon}\right)^{x} \ldots\ldots\ldots\ldots (17),$$

the ratio of the two members tending to unity as x tends to infinity. And speaking generally it is with the ratios, not the actual values of functions of large numbers, that we are concerned.

6. Yet even in cases in which the value of x involved is not only large but infinite, it may be necessary to take account of terms involving negative powers; just as in the

Differential Calculus it is sometimes necessary to proceed to the higher orders of differential coefficients.

Ex. 5.   Required an approximate value of the fraction

$$\frac{2\cdot4\dots2x}{3\cdot5\dots(2x-1)},$$

x being large, but account being taken of terms of the order x^{-1}.

We have, by the last example, under the conditions specified,

$$1\cdot2\cdot3\dots\ x=\sqrt{(2\pi)}\,x^{x+\frac12}\,\epsilon^{-x}\left(1+\frac{1}{12x}\dots\right),$$

$$2\cdot4\cdot6\dots2x=2^{x+\frac12}\sqrt{\pi}\,x^{x+\frac12}\,\epsilon^{-x}\left(1+\frac{1}{12x}\right),$$

$$1\cdot2\cdot3\dots2x=\sqrt{(2\pi)}\,(2x)^{2x+\frac12}\,\epsilon^{-2x}\left(1+\frac{1}{24x}\right),$$

$$3\cdot5\dots(2x-1)=2^{x+\frac12}x^x\epsilon^{-x}\left(\frac{1+\dfrac{1}{24x}}{1+\dfrac{1}{12x}}\right)$$

$$=2^{x+\frac12}x^x\epsilon^{-x}\left(1-\frac{1}{24x}\dots\right),$$

on developing the fraction by division.

Hence

$$\frac{2\cdot4\dots2x}{3\cdot5\dots(2x-1)}=\pi^{\frac12}x^{\frac12}\left(\frac{1+\dfrac{1}{12x}}{1-\dfrac{1}{24x}}\right)$$

$$=\pi^{\frac12}x^{\frac12}\left(1+\frac{1}{8x}\dots\right)\dots\dots(18).$$

Ex. 6.   To sum the series

$$1+\frac{1}{2^{2n}}+\frac{1}{3^{2n}}+\frac{1}{4^{2n}}\dots+\frac{1}{x^{2n}}.$$

Representing the series by u, we have

$$u = \frac{1}{x^{2n}} + \Sigma \frac{1}{x^{2n}}$$

$$= C - \frac{1}{(2n-1)\, x^{2n-1}} + \frac{1}{2x^{2n}} - \frac{2n}{12x^{2n+1}}$$

$$+ \frac{2n\,(2n+1)\,(2n+2)}{720x^{2n+3}}, \ \&c.$$

For each particular value of n the constant C might be determined approximately as in Ex. 3, but its general expression may be found as follows.

Let x become infinite, then the above equation gives

$$1 + \frac{1}{2^{2n}} + \frac{1}{3^{2n}} + \&c. \ ad\ inf. = C,$$

and it remains to evaluate the series in the first member.

Now

$$\cot \theta = \sqrt{(-1)}\, \frac{\epsilon^{2\theta\sqrt{(-1)}} + 1}{\epsilon^{2\theta\sqrt{(-1)}} - 1} = \sqrt{(-1)} \left(1 + \frac{2}{\epsilon^{2\theta\sqrt{(-1)}} - 1} \right)$$

$$= \sqrt{(-1)} \left[1 + 2 \left\{ \frac{1}{2\theta\sqrt{(-1)}} - \frac{1}{2} + \frac{B_1\, 2\theta\,\sqrt{(-1)}}{1.2} - \&c. \right\} \right]$$

$$= \frac{1}{\theta} - \frac{2^2}{1.2}\, B_1\, \theta - \frac{2^4\, B_3}{1.2.3.4}\, \theta^3 - \&c.$$

in which the coefficient of θ^{2n-1} is $\dfrac{-\, 2^{2n}\, B_{2n-1}}{1.2 \ldots 2n}$.

Again, by a known theorem,

$$\sin \theta = \theta \left(1 - \frac{\theta^2}{\pi^2} \right) \left(1 - \frac{\theta^2}{2^2 \pi^2} \right) \left(1 - \frac{\theta^2}{3^2 \pi^2} \right) \ldots\ldots$$

Therefore, taking the logarithm of each member and differentiating,

$$\cot\theta = \frac{1}{\theta} - \frac{2\theta}{\pi^2}\left(1 - \frac{\theta^2}{\pi^2}\right)^{-1} - \frac{2\theta}{2^2\pi^2}\left(1 - \frac{\theta^2}{2^2\pi^2}\right)^{-1}$$

$$- \frac{2\theta}{3^2\pi^2}\left(1 - \frac{\theta^2}{3^2\pi^2}\right)^{-1} - \&c.$$

$$= \frac{1}{\theta} - \frac{2}{\pi^2}\left(1 + \frac{1}{2^2} + \frac{1}{3^2} + \&c.\right)\theta - \frac{2}{\pi^4}\left(1 + \frac{1}{2^4} + \frac{1}{3^4} + \&c.\right)\theta^3$$

$$- \frac{2}{\pi^6}\left(1 + \frac{1}{2^6} + \frac{1}{3^6} + \&c.\right)\theta^5 + \&c.,$$

in which the coefficient of θ^{2n-1} is

$$\frac{-2}{\pi^{2n}}\left(1 + \frac{1}{2^{2n}} + \frac{1}{3^{2n}} + \&c.\right).$$

Equating this to the value before obtained for the same element, we have

$$1 + \frac{1}{2^{2n}} + \frac{1}{3^{2n}} + \frac{1}{4^{2n}} + \&c. = \frac{(2\pi)^{2n}B_{2n-1}}{2\,(1\,.\,2\,\ldots\,2n)} \quad\ldots\ldots\ (19).$$

Hence

$$u = \frac{(2\pi)^{2n}}{2\,(1\,.\,2\,\ldots\,2n)} - \frac{1}{(2n-1)\,x^{2n-1}} + \frac{1}{2x^{2n}} - \frac{n}{6x^{2n+1}}$$

$$+ \frac{2n\,(2n+1)\,(2n+2)}{720x^{2n+3}} + \&c.\ldots\ldots(20),$$

the expression required.

7.   The above investigation gives us also another expression for Bernoulli's numbers, viz.

$$B_{2n-1} = \frac{2\,(1\,.\,2\,\ldots\,2n)}{(2\pi)^{2n}}\left(1 + \frac{1}{2^{2n}} + \frac{1}{3^{2n}} + \frac{1}{4^{2n}} + \&c.\right)\ldots(21),$$

while for the *divided* coefficients which appear in the actual expansion of Σu_x we have the expression

$$\frac{B_{2n-1}}{1\,.\,2\,\ldots\,2n} = \frac{2}{(2\pi)^{2n}}\left(1 + \frac{1}{2^{2n}} + \frac{1}{3^{2n}} + \&c.\right)\quad\ldots\ldots\ (22).$$

Hence too it appears, First, that the series of Bernoulli's numbers is ultimately divergent, for, n becoming indefinitely great, we have

$$\text{Limit of } \frac{B_{2n+1}}{B_{2n-1}} = \frac{n^2}{\pi^2} \dots\dots\dots\dots (23).$$

Secondly, that the *divided* coefficients in the expansion of Σu_x are ultimately convergent. For

$$\text{Limit} \left\{ \frac{B_{2n+1}}{1 \cdot 2 \dots (2n+2)} \div \frac{B_{2n-1}}{1 \cdot 2 \dots 2n} \right\} = \frac{2}{(2\pi)^{2n+2}} \div \frac{2}{(2\pi)^{2n}}$$

$$= \frac{1}{4\pi^2} \dots\dots\dots\dots (24).$$

The expansion itself will be ultimately convergent or divergent according to circumstances.

Limits of the Remainder of the Series for Σu_x.

8. Representing, for simplicity, u_x by u, we have

$$\Sigma u = C + \int u \, dx - \tfrac{1}{2} u + \frac{B_1}{1 \cdot 2} \frac{du}{dx} \dots + (-1)^{n-1} \frac{B_{2n-1}}{1 \cdot 2 \dots 2n} \frac{d^{2n-1} u}{dx^{2n-1}}$$

$$+ \Sigma_{r=n+1}^{r=\infty} (-1)^{r-1} \frac{B_{2r-1}}{1 \cdot 2 \dots 2r} \frac{d^{2r-1} u}{dx^{2r-1}} \dots\dots\dots\dots (25).$$

The second line of this expression we shall represent by R, and endeavour to determine the limits of its value.

Now by (22),

$$\frac{B_{2r-1}}{1 \cdot 2 \dots 2r} = \frac{2}{(2\pi)^{2r}} \Sigma_{m=1}^{m=\infty} \frac{1}{m^{2r}} .$$

Therefore substituting,

$$R = \Sigma_{r=n+1}^{r=\infty} \Sigma_{m=1}^{m=\infty} \frac{2 \, (-1)^{r-1}}{(2\pi)^{2r} m^{2r}} \frac{d^{2r-1} u}{dx^{2r-1}}$$

$$= 2 \Sigma_{m=1}^{m=\infty} \Sigma_{r=n+1}^{r=\infty} \frac{(-1)^{r-1}}{(2m\pi)^{2r}} \frac{d^{2r-1} u}{dx^{2r-1}} .$$

Assume

$$\Sigma_{r=n+1}^{r=\infty} \frac{(-1)^{r-1}}{(2m\pi)^{2r}} \frac{d^{2r-1} u}{dx^{2r-1}} = t.$$

And then, making $\dfrac{1}{2m\pi} = \epsilon^\theta$, we are led by the general theorem for the summation of series (*Diff. Equations*, p. 431)[*] to the differential equation

$$t + \frac{d^2}{dx^2}\epsilon^{2\theta}t = (-1)^n \frac{d^{2n+1}u}{dx^{2n+1}}\epsilon^{(2n+2)\theta},$$

$$\text{or } \frac{d^2t}{dx^2} + (2m\pi)^2 t = \frac{(-1)^n}{(2m\pi)^{2n}}\frac{d^{2n+1}u}{dx^{2n+1}},$$

the complete integral of which is (*Diff. Equations*, p. 383)

$$t = \frac{(-1)^n}{(2m\pi)^{2n+1}}\left\{\sin 2m\pi x \int\cos 2m\pi x \frac{d^{2n+1}u}{dx^{2n+1}}\,dx\right.$$
$$\left. - \cos 2m\pi x \int\sin 2m\pi x \frac{d^{2n+1}u}{dx^{2n+1}}\,dx\right\},$$

or, since we have to do only with integer values of x for which $\sin(2m\pi x) = 0$, $\cos(2m\pi x) = 1$,

$$t = \frac{(-1)^{n+1}}{(2m\pi)^{2n+1}}\int\sin 2m\pi x \frac{d^{2n+1}u}{dx^{2n+1}}\,dx.$$

Hence

$$R = 2\sum_{m=1}^{m=\infty}\frac{(-1)^{n+1}}{(2m\pi)^{2n+1}}\int\sin 2m\pi x \frac{d^{2n+1}u}{dx^{2n+1}}\,dx$$
$$= 2(-1)^{n+1}\int\left\{\frac{\sin 2\pi x}{(2\pi)^{2n+1}} + \frac{\sin 4\pi x}{(4\pi)^{2n+1}} + \&c.\right\}\frac{d^{2n+1}u}{dx^{2n+1}}\,dx \ldots(26),$$

the lower limit of integration being such a value of x as makes $\dfrac{d^{2n+1}u}{dx^{2n+1}}$ to vanish, the upper limit x. Hence if *within*

---

[*] The theorem which is of most frequent application is the following. If $u = u_p x^p + u_{p+r} x^{p+r} + u_{p+2r} x^{p+2r} + \&c.$ *ad inf.*, then supposing the law of the coefficients to be

$$u_m = \phi(m) u_{m-r}, \text{ or } u_m - \phi(m) u_{m-r} = 0,$$

and making $x = \epsilon^\theta$, the differential equation for u will be

$$u - \phi\left(\frac{d}{d\theta}\right)\epsilon^{r\theta} = u_p \epsilon^{p\theta}.$$

The student will easily deduce this from the more general theorem referred to.

the limits of integration $\dfrac{d^{2n+1}u}{dx^{2n+1}}$ retain a constant sign, the value of R will be *numerically* less than that of the function

$$2 \int \left\{ \frac{1}{(2\pi)^{2n+1}} + \frac{1}{(4\pi)^{2n+1}} \cdots \right\} \frac{d^{2n+1}u}{dx^{2n+1}} \, dx \, ;$$

therefore, than that of the function

$$2 \left\{ \frac{1}{(2\pi)^{2n+1}} + \frac{1}{(4\pi)^{2n+1}} \cdots \text{ ad inf.} \right\} \frac{d^{2n}u}{dx^{2n}} \, ;$$

therefore, by (22), than that of the function

$$\frac{1}{2\pi} \frac{B_{2n-1}}{1 . 2 \ldots 2n} \frac{d^{2n}u}{dx^{2n}} \quad \cdots\cdots\cdots\cdots\cdots\cdots (27).$$

When n is large this expression tends to a strict interpolation of *form* between the last term of the series given and the first term of its remainder, viz., omitting signs, between

$$\frac{B_{2n-1}}{1 . 2 \ldots 2n} \frac{d^{2n-1}u}{dx^{2n-1}} \text{ and } \frac{B_{2n+1}}{1 . 2 \ldots (2n+2)} \frac{d^{2n+1}u}{dx^{2n+1}} \cdots\cdots(28),$$

it being remembered that by (24) the coefficient of $\dfrac{d^{2n}u}{dx^{2n}}$ in (26) is, in the *limit*, a mean proportional between the coefficients of $\dfrac{d^{2n-1}u}{dx^{2n-1}}$ and $\dfrac{d^{2n+1}u}{dx^{2n+1}}$ in (28). And this interpolation of form is usually accompanied by interpolation of value, though without specifying the form of the function u we can never affirm that such will be the case.

The practical conclusion is that the summation of the convergent terms of the series for Σu affords a sufficient approximation, except when the first differential coefficient in the remainder changes sign within the limits of integration.

The series for

$$\Sigma \log x, \quad \Sigma \frac{1}{x^m}, \quad \Sigma \frac{1}{(ax+b)^m},$$

are therefore of the kind in which the summation of convergent terms suffices.

Particular forms of Σu_x.

9.   Beside the general expression for Σu_x given in Art. 2, there are certain other forms which suppose a particular constitution of the function u_x, and are advantageous under particular circumstances.

Thus the series

$$\phi(0) - \phi(1) + \phi(2) - \phi(3) + \&c.$$

has for its $x+1^{\text{th}}$ term $(-1)^x \phi(x)$, and for the sum of its first x terms $\Sigma(-1)^x \phi(x)$. Applying the theorem of Art. 2, we should obtain for this an expansion proceeding according to the differential coefficients of $(-1)^x \phi(x)$. It is obviously desirable to substitute for this an expansion proceeding according to the differential coefficients of $\phi(x)$.

By a theorem employed in the demonstration of the fourth integrable form of Chap. IV. Art. 2, we have

$$\Sigma(-1)^x \phi(x) = (-1)^x (-\epsilon^{\frac{d}{dx}} - 1)^{-1} \phi(x)$$

$$= (-1)^{x-1} (\epsilon^{\frac{d}{dx}} + 1)^{-1} \phi(x).$$

Now

$$\frac{1}{\epsilon^t + 1} = \frac{1}{\epsilon^t - 1} - 2\left(\frac{1}{\epsilon^{2t} - 1}\right)$$

$$= \frac{1}{t} - \frac{1}{2} + \frac{B_1}{1.2} t - \frac{B_3}{1.2.3.4} t^3 + \&c.$$

$$- 2\left\{\frac{1}{2t} - \frac{1}{2} + \frac{B_1}{1.2} 2t - \frac{B_3}{1.2.3.4} (2t)^3 + \&c.\right\}$$

$$= \frac{1}{2} + \frac{B_1}{1.2}(1 - 2^2) t - \frac{B_3}{1.2.3.4}(1 - 2^4) t^3 + \&c.,$$

whence

$$(\epsilon^{\frac{d}{dx}} - 1)^{-1} \phi(x) = \left\{\frac{1}{2} + \frac{B_1}{1.2}(1 - 2^2)\frac{d}{dx}\right.$$

$$\left. - \frac{B_3}{1.2.3.4}(1 - 2^4)\frac{d^3}{dx^3} + \&c.\right\} \phi(x).$$

Therefore

$$\Sigma \, (-1)^x \, \phi \, (x) = C + (-1)^x \left\{ - \frac{1}{2} \, \phi \, (x) + \frac{B_1}{1 \cdot 2} \, (2^2 - 1) \, \phi' \, (x) \right.$$

$$\left. - \frac{B_3}{1 \cdot 2 \cdot 3 \cdot 4} \, (2^4 - 1) \, \phi''' \, (x) + \frac{B_5}{1 \cdot 2 \dots 6} \, (2^6 - 1) \, \phi^v \, (x) - \&c. \right\}$$
$$\dots\dots (29).$$

Or, on calculating a few of the coefficients,

$$\Sigma \, (-1)^x \, \phi \, (x) = C + (-1)^x \left\{ - \frac{1}{2} \, \phi \, (x) + \frac{1}{4} \, \phi' \, (x) \right.$$

$$\left. - \frac{1}{48} \, \phi''' \, (x) + \frac{1}{480} \, \phi^v \, (x) - \frac{17}{80640} \, \phi^{vii} \, (x) + \&c. \right\}$$
$$\dots\dots (30),$$

the theorem required.

10. It is possible also to develope $\Sigma \, (-1)^x \, \phi \, (x)$ in ascending *differences* of $\phi \, (x)$.

We have by Art. 9,

$$\Sigma \, (-1)^x \, \phi \, (x) = (-1)^{x-1} \, (\epsilon^{\frac{d}{dx}} + 1)^{-1} \, \phi \, (x)$$

$$= (-1)^{x-1} \, (2 + \Delta)^{-1} \, \phi \, (x)$$

$$= C + (-1)^{x-1} \left\{ \frac{\phi \, (x)}{2} - \frac{\Delta \phi \, (x)}{4} + \frac{\Delta^2 \phi \, (x)}{8} - \&c. \right\}$$
$$\dots\dots (31).$$

Suppose it required to deduce hence an expression for the series

$$\phi \, (0) - \phi \, (1) + \phi \, (2) - \phi \, (3) + \&c. \; ad \; inf.,$$

the terms being supposed to converge to 0. In (31) let x be successively made 0 and ∞, we have

$$0 = C - \frac{\phi \, (0)}{2} + \frac{\Delta \phi \, (0)}{4} - \&c.$$

$$\phi \, (0) - \phi \, (1) + \phi \, (2) - \&c. \; ad \; inf. = C.$$

Hence

$$\phi(0) - \phi(1) + \phi(2) - \&c. \ ad \ inf.$$

$$= \frac{\phi(0)}{2} - \frac{\Delta\phi(0)}{4} + \frac{\Delta^2\phi(0)}{8} - \&c. \ \ldots\ldots\ldots (32).$$

This theorem, which may be applied with great advantage to the summation of slowly convergent series whose terms are alternately positive and negative, may also be established as follows,

$$\phi(0) - \phi(1) + \phi(2) - \&c. = (1 - D + D^2 - D^3 + \&c.)\ \phi(0)$$

$$= (1 + D)^{-1}\ \phi(0)$$

$$= (2 + \Delta)^{-1}\ \phi(0)$$

$$= \left(\frac{1}{2} - \frac{\Delta}{4} + \frac{\Delta^2}{8} - \frac{\Delta^3}{16} + \&c.\right)\ \phi(0)$$

$$= \frac{\phi(0)}{2} - \frac{\Delta\phi(0)}{4} + \frac{\Delta^2\phi(0)}{8} - \&c.$$

11. The above results are virtually included in the two following more general theorems by which $\Sigma\phi(x)t^x$ is expressed in series proceeding according to the differential coefficients and according to the differences of $\phi(x)$, viz.

1st. $$\Sigma\phi(x)t^x = C - \frac{t^x}{1 - t}\left\{\phi(x) + A_1\frac{d\phi(x)}{dx}\right.$$

$$\left. + \frac{A_2}{1.2}\frac{d^2\phi(x)}{dx^2} + \&c.\right\} \ \ldots\ldots(33),$$

where $$A_n = \left(1 - \frac{t}{1-t}\Delta\right)^{-1}0^n$$

$$= \frac{t}{1-t}\Delta 0^n + \left(\frac{t}{1-t}\right)^2\Delta^2 0^n + \left(\frac{t}{1-t}\right)^3\Delta^3 0^n + \&c.$$

2ndly. $$\Sigma\phi(x)t^x = C - \frac{t^x}{1-t}\left\{\phi(x) + \frac{t}{1-t}\Delta\phi(x)\right.$$

$$\left. + \left(\frac{t}{1-t}\right)^2\Delta^2\phi(x) + \&c.\right\} \ \ldots\ldots (34).$$

To demonstrate these theorems we must write, Chap. IV. p. 53,

$$\Sigma \phi(x) t^x = t^x (te^{\frac{d}{dx}} - 1)^{-1} \phi(x),$$
$$= t^x \{t(1 + \Delta) - 1\}^{-1} \phi(x),$$

and develope the symbolical functions in the last two members in ascending powers of $\frac{d}{dx}$ and of Δ respectively, the former expansion being effected by the reciprocal form of Maclaurin's theorem, Chap. II. Art. 11.

EXAMPLES.

1. Approximate to the numerical value of the series

$$\frac{1}{1} + \frac{1}{2} + \frac{1}{3} \cdots + \frac{1}{1000}.$$

2. Deduce an approximate value of $\log 1 + \log 2 + \log 3 \cdots + \log 1000$ in the Napierian system.

3. If an approximate value for the series

$$1 + \frac{1}{2} + \frac{1}{3} \cdots + \frac{1}{x}$$

be expressed in the form

$$C + \log x + \frac{1}{2x} - \frac{B_1}{2x^2} + \frac{B_3}{4x^4} - \frac{B_5}{6x^6} + \&c.$$

shew that

$$C = \frac{1}{2} + \frac{B_1}{2} - \frac{B_3}{4} + \frac{B_5}{6} - \&c.,$$

and hence compute an approximate value of C.

4. Find an approximation for

$$\frac{3 \cdot 5 \cdots (2x + 1)}{2 \cdot 4 \cdot 2x},$$

B. F. D.

7

supposing x large; the first negative power of x which presents itself being retained.

5. Transform the series

$$1 - \frac{1}{2} + \frac{1}{3} - \frac{1}{4} + \frac{1}{5} - \&\text{c.}$$

to one of a more convergent character.

6. The series $a_1 x - a_2 x^2 + a_3 x^3 - \&\text{c.}$ may be transformed into

$$\frac{a_1 x}{1 + x} - \Delta a_1 \frac{x^2}{(1 + x)^2} + \Delta^2 a_1 \frac{x^3}{(1 + x)^3} - \&\text{c.},$$

where
$$\Delta a_n = a_{n+1} - a_n.$$

CHAPTER VII.

EQUATIONS OF DIFFERENCES.

1. An ordinary equation of differences is an expressed relation between an independent variable x, a dependent variable u_x, and any successive differences of u_x, as Δu_x, $\Delta^2 u_x$ ...$\Delta^n u_x$. The order of the equation is determined by the order of its highest difference; its degree by the index of the power in which that highest difference is involved, supposing the equation rational and integral in form.

Equations of differences may also be presented in a form involving successive values, instead of successive differences, of the dependent variable; for $\Delta^n u_x$ can be expressed in terms of u_x, $u_{x+1} \ldots u_{x+n}$. Chap. II. Art. 10.

Equations of differences are said to be linear when, considered functionally, they are of the first degree with respect to u_x, Δu_x, $\Delta^2 u_x$, &c.; or, supposing successive values of the independent variable to be employed instead of successive differences, when they are of the first degree with respect to u_x, u_{x+1}, u_{x+2}, &c. The connexion of the two cases is obvious.

Equations of differences which do not belong to the ordinary species, viz. equations of partial differences and of mixed differences, will be defined in another chapter.

Genesis of Equations of Differences.

2. The genesis of equations of differences is analogous to that of differential equations. From a *complete* primitive

$$F(x, u_x, c) = 0 \ \ldots\ldots\ldots\ldots\ldots\ldots (1),$$

connecting a dependent variable u_x with an independent

variable x and an arbitrary constant c, and from the derived equation

$$\Delta F(x, u_x, c) = 0 \quad\dotfill\quad (2),$$

we obtain, by eliminating c, an equation of the form

$$\phi(x, u_x, \Delta u_x) = 0 \quad\dotfill\quad (3).$$

Or, if successive values are employed in the place of differences, an equation of the form

$$\psi(x, u_x, u_{x+1}) = 0 \quad\dotfill\quad (4).$$

Either of these may be considered as a type of equations of differences of the first order.

In like manner if, from a complete primitive

$$F(x, u_x, c_1, c_2, \dots c_n) = 0 \quad\dotfill\quad (5),$$

and from n successive equations derived from it by successive performances of the operation denoted by Δ, we eliminate $c_1, c_2 \dots c_n$, we obtain an equation which will assume the form

$$\phi(x, u_x, \Delta u_x, \dots \Delta^n_{\mu} u_x) = 0 \quad\dotfill\quad (6),$$

or the form

$$\psi(x, u_x, u_{x+1}, \dots u_{x+n}) = 0 \quad\dotfill\quad (7),$$

according as successive differences or successive values are employed. Either of these forms is typical of equations of differences of the nth order. In (6), u_x may be replaced by u.

Ex. 1. Assuming as complete primitive $u = cx + c^2$, we have, on taking the difference,

$$\Delta u = c,$$

by which, eliminating c, there results

$$u = x\Delta u + (\Delta u)^2,$$

the corresponding equation of differences of the first order.

Thus too for any complete primitive of the form $u = cx + f(c)$ there will exist an equation of differences of the form

$$u = x\Delta u + f(\Delta u) \quad\dotfill\quad (8).$$

Ex. 2. Assuming as complete primitive

$$u_x = ca^x + c'b^x,$$

we have

$$u_{x+1} = ca^{x+1} + c'b^{x+1},$$

$$u_{x+2} = ca^{x+2} + c'b^{x+2}.$$

Hence

$$u_{x+1} - au_x = c'(b - a)\, b^x,$$

$$u_{x+2} - au_{x+1} = c'(b - a)\, b^{x+1}.$$

Therefore

$$u_{x+2} - au_{x+1} - b\,(u_{x+1} - au_x) = 0,$$

or

$$u_{x+2} - (a + b)\, u_{x+1} + abu_x = 0 \dots\dots\dots (9).$$

Here two arbitrary constants being contained in the complete primitive, the equation of differences is of the second order.

3. The arbitrary constants in the complete primitive of an equation of differences are properly speaking periodical functions of x of the kind whose nature has been explained, and whose analytical expression has been determined in Chap. IV. Art. 1. They are constant with reference only to the operation Δ, and as such, are subject only to the condition of resuming the same value for values of x differing by unity; a condition which however reduces them to absolute constants when x admits of integral values only. The subject will be more fully discussed in Arts. 4 and 11 of this chapter.

The proposition converse to that of Art. 2, viz. that a complete primitive of an equation of differences of the nth order involves n arbitrary constants, has already been established by general considerations (*Diff. Equations*, p. 187).

Linear Equations of the First Order.

4. The typical form of this class of equations is

$$u_{x+1} - A_x u_x = B_x \dots\dots\dots\dots\dots (1),$$

where A_x and B_x are given functions of x. We shall first consider the case in which the second member is 0.

To integrate the equation

$$u_{x+1} - A_x u_x = 0 \dots\dots\dots\dots\dots (2),$$

we have
$$u_{x+1} = A_x u_x,$$
whence, the equation being true for all values of x,
$$u_x = A_{x-1} u_{x-1},$$
$$u_{x-1} = A_{x-2} u_{x-2},$$
$$\dots\dots\dots\dots\dots$$
$$u_{r+1} = A_r u_r.$$
Hence, by successive substitutions,
$$u_{x+1} = A_x A_{x-1} A_{x-2} \dots A_r u_r \dots\dots\dots (3),$$
r being an assumed initial value of x.

Let C be the arbitrary value of u_x corresponding to $x = r$, (arbitrary because it being fixed the succeeding values of u_x, corresponding to $x = r+1$, $x = r+2$, &c., are determined in succession by (2), while u_r is itself left undetermined) then (3) gives
$$u_{x+1} = C A_x A_{x-1} \dots A_r,$$
whence
$$u_x = C A_{x-1} A_{x-2} \dots A_r, \dots\dots\dots(4),$$
and this is the general integral sought.

While, for any particular system of values of x differing by successive unities, C is an arbitrary constant, for the aggregate of all possible systems it is a periodical function of x, whose cycle of change is completed while x varies continuously *through* unity. Thus, suppose the initial value of x to be 0, then whatever arbitrary value we assign to u_0, the values of u_1, u_2, u_3, &c. are rigorously determined by the equation (2). Here then C, which represents the value of u_0, is an arbitrary constant, and we have
$$u_{x+1} = C A_x A_{x-1} \dots A_0.$$
Suppose however the initial value of x to be $\frac{1}{2}$, and let E be the corresponding value of u_x. Then, whatever arbitrary value we assign to E, the system of values of $u_{\frac{3}{2}}$, $u_{\frac{5}{2}}$, &c. will be rigorously determined by (2), and the solution becomes
$$u_{x+1} = E A_x A_{x-1} \dots A_{\frac{1}{2}}.$$

The given equation of differences establishes however no connexion between C and E. *The aggregate of possible solutions is therefore comprised in* (4) *supposing C therein to be an arbitrary periodical function of x completing its changes while x changes through unity, and therefore becoming a simple arbitrary constant for any system of values of x differing by successive unities.*

We may for convenience express (4) in the form

$$u_x = CPA_{x-1} \dots\dots\dots\dots (5),$$

where P is a symbol of operation denoting the indefinite continued product of the successive values which the function of x, which it precedes, assumes while x successively decreases by unity.

There is another mode of deducing this result which it may be well to notice.

Let $u_x = \epsilon^t$. Then $u_{x+1} = \epsilon^{t+\Delta t}$, and (2) becomes

$$\epsilon^{t+\Delta t} - A_x \epsilon^t = 0;$$

$$\therefore\ \epsilon^{\Delta t} - A_x = 0,$$

whence $\quad \Delta t = \log A_x,$

$$t = \Sigma \log A_x + C,$$

$$= \log A_{x-1} + \log A_{x-2} + \&c. + C,$$

$$= \log PA_{x-1} + C.$$

Therefore

$$u_x = \epsilon^{\log PA_{x-1}+C}, = C_1 PA_{x-1},$$

as before.

Resuming the general equation (1) let us give to u_x the form above determined, only replacing C by a variable parameter C_x, and then, in analogy with the known method of solution for linear differential equations, seek to determine C_x.

We have $\quad u_x = C_x PA_{x-1},$

$$u_{x+1} = C_{x+1} PA_x,$$

whence (1) becomes

$$C_{x+1} PA_x - A_x C_x PA_{x-1} = B_x.$$

But $\qquad A_x PA_{x-1} = PA_x,$

whence $\qquad (C_{x+1} - C_x) PA_x = B_x,$

or, $\qquad (\Delta C_x) PA_x = B_x,$

whence $\qquad \Delta C_x = \dfrac{B_x}{PA_x},$

$$C_x = \Sigma \frac{B_x}{PA_x} + C \dots\dots\dots\dots (6);$$

$$\therefore u_x = PA_{x-1} \left\{ \Sigma \frac{B_x}{PA_x} + C \right\} \dots\dots\dots (7),$$

the general integral sought.

Ex. 1. Given $u_{x+1} - (x+1) u_x = 1 . 2 \dots (x+1).$

From the form of the second member it is apparent that x admits of integral values only.

Here $\qquad A_x = x + 1, \quad PA_{x-1} = x(x-1) \dots 1,$

$$\frac{B_x}{PA_x} = 1, \quad \Sigma \frac{B_x}{PA_x} = x;$$

$$\therefore u_x = x(x-1) \dots 1 \times (x + C).$$

Ex. 2. Given $u_{x+1} - au_x = b$, where a and b are constant.

Here $\qquad A_x = a,$ and $PA_x = a^x,$ therefore

$$u_x = a^{x-1} \left\{ \Sigma \frac{b}{a^x} + C \right\}$$

$$= \frac{b}{1-a} + C_1 a^x,$$

where C_1 is an arbitrary constant.

We may observe before dismissing the above example, that when $A_x = a$ the complete value of PA_x is a^x multiplied by an indeterminate constant. For

$$PA_x = A_x A_{x-1} \dots A_r$$
$$= a . a . a \dots , \quad x - r + 1 \text{ times,}$$
$$= a^{x-r+1} = a^{-r+1} \times a^x.$$

But were this value employed, the indeterminate constant a^{-r+1} would in one term of the general solution (7) disappear by division, and in the other merge into the arbitrary constant C. Actually we made use of the particular value corresponding to $r = 1$, and this is what in most cases it will be convenient to do.

Ex. 3.   Given $\Delta u_x + 2u_x = -x - 1.$

Replacing Δu_x by $u_{x+1} - u_x$, we have

$$u_{x+1} + u_x = -(x+1).$$

Here $A_x = -1$, $B_x = -(x+1)$, whence substituting in (7) and reducing

$$u_x = C(-1)^x - \frac{x}{2} - \frac{1}{4}.$$

Ex. 4. $u_{x+1} - au_x = \dfrac{a^x}{(x+1)^2}.$

We find

$$u_x = a^{x-1} \left\{ \Sigma \frac{1}{(x+1)^2} + C \right\}$$

$$= a^{x-1} \left\{ \frac{1}{1^2} + \frac{1}{2^2} \dots + \frac{1}{x^2} + C \right\}.$$

When, as in the above example, the summation denoted by Σ cannot be effected in *finite* terms, it is convenient to employ as above an indeterminate series. In so doing we have supposed the solution to have reference to positive and integral values of x. The more general form would be

$$u_x = a^{x-1} \left\{ \frac{1}{x^2} + \frac{1}{(x-1)^2} \dots + \frac{1}{r^2} \right\},$$

r being the initial value of x.

Linear Equations with constant Coefficients.

5.   The general form of a linear equation with constant coefficients when expressed by successive *values* not increments of the independent variable, and having its second member equal to 0, is

$$u_{x+n} + A_1 u_{x+n-1} + A_2 u_{x+n-2} \dots + A_n u_x = 0.$$

If we assume $u_x = cm^x$, we obtain, on substitution and division by cm^x, the 'auxiliary equation'

$$m^n + A_1 m^{n-1} + A_2 m^{n-2} \dots + A_n = 0.$$

There exist therefore, when the roots of this equation are all different, n particular solutions of the form assumed above, corresponding to the n particular determinations of m.   It is also evident from the linear form of the given equation, that the *general* value of u_x will be the sum of the n particular values thus obtained.

Ex. 5.   The equation $u_{x+2} - 5u_{x+1} + 6u_x = 0$ leads to the auxiliary equation

$$m^2 - 5m + 6 = 0,$$

and therefore admits of the two particular integrals

$$u_x = c_1 (2)^x, \quad u_x = c_2 (3)^x,$$

and of the general integral

$$u_x = c_1 (2)^x + c_2 (3)^x.$$

When the auxiliary equation has imaginary roots, or equal roots, or when the given equation has a second member, the principles employed in the corresponding cases of differential equations may be adopted without essential change.

If m have imaginary values, their trigonometrical equivalents must be employed.

Ex. 6.   Given $u_{x+2} + a^2 u_x = 0.$

Here $u_x = c_1 \{a \sqrt{(-1)}\}^x + c_2 \{-a \sqrt{(-1)}\}^x$

$$= a^x \left[c_1 \left\{\cos \frac{\pi}{2} + \sqrt{(-1)} \sin \frac{\pi}{2}\right\}^x + c_2 \left\{\cos \frac{\pi}{2} - \sqrt{(-1)} \sin \frac{\pi}{2}\right\}^x \right]$$

$$= a^x \left\{A \cos \left(\frac{\pi x}{2}\right) + B \sin \left(\frac{\pi x}{2}\right)\right\}.$$

If m have equal roots the solution may be derived by the method of limits (assuming the principle of continuity) from the answering form of the solution when the roots are unequal.

Ex. 7.   Given $u_{x+2} - 2au_{x+1} + a^2 u_x = 0$.

The solution is the limiting form of

$$u_x = c_1 a^x + c_2 b^x,$$

b approaching to a.   Therefore

$$u_x = \text{lim of} \left(Ca^x + C' \frac{b^x - a^x}{b - a} \right)$$

$$= Ca^x + C'xa^{x-1} = a^x \left(C + C_1 x \right).$$

If the second member be not equal to 0 the form of the solution may be deduced from the particular form which it assumes when the second member is 0, by treating the constants as variable parameters.

But in the last case, and usually in the preceding one, it is simpler to proceed by the symbolical method of the following section.

Symbolical solution of equations with constant coefficients.

6.   The linear equation of the nth order with constant coefficients is of the form

$$u_{x+n} + A_1 u_{x+n-1} + A_2 u_{x+n-2} \dots + A_n u_x = X \dots\dots\dots(1).$$

Hence if D be a symbol of operation defined by

$$D\phi(x) = \phi(x+1) \dots\dots\dots\dots\dots(2),$$

the above linear equation becomes

$$D^n u_x + A_1 D^{n-1} u_x + A_2 D^{n-2} u_x \dots + A_n u_x = X,$$

or, separating the symbols of operation,

$$(D^n + A_1 D^{n-1} + A_2 D^{n-2} \dots + A_n) u_x = X,$$

and this we shall for brevity represent by

$$f(D) u = X \dots\dots\dots\dots\dots\dots(3);$$

whence $\qquad u = \{f(D)\}^{-1} X$ ....(4),

a form indeed differing in interpretation from (3) only in that it presents u as the object of quest (*Diff. Equations*, p. 375). Suppose the roots of the auxiliary equation $f(m) = 0$ to be all different, then, resolving $\{f(D)\}^{-1}$ as if it were a rational fraction, we have a result of the form

$$u = \{N_1(D - a_1)^{-1} + N_2(D - a_2)^{-1} \ldots + N_n(D - a_n)^{-1}\} X * \ldots(5),$$

where $\qquad N_1 = \dfrac{1}{(a_1 - a_2)(a_1 - a_3)\ldots(a_1 - a_n)},$

with similar expressions for $N_2 \ldots N_n$. But if a particular root a be repeated r times, then we have corresponding to that root a series of terms

$$\{M_1(D - a)^{-1} + M_2(D - a)^{-2} \ldots + M_r(D - a)^{-r}\} X \ldots(6),$$

where $\qquad M_i = \dfrac{1}{1.2\ldots(r-i)} \left(\dfrac{d}{dz}\right)^{r-i} \dfrac{f(z)}{(z - a)^r}, \quad (z = a).$

The solution of the proposed equation is therefore made to depend in all cases on the performance of the operation denoted by $(D - a)^{-i} X$; and this, since we are permitted by the symbolical form of Taylor's theorem to substitute $\epsilon^{\frac{d}{dx}}$ for D, may be referred to the known theorems

$$\phi\left(\dfrac{d}{dx}\right) \epsilon^{mx} = \phi(m) \epsilon^{mx}, \quad \phi\left(\dfrac{d}{dx}\right) \epsilon^{mx} X = \epsilon^{mx} \phi\left(\dfrac{d}{dx} + m\right) X \ldots(7).$$

(*Diff. Equations*, p. 384). Let us first suppose the subject of the operation to be $a^x X$ where X denotes any function of x. Then

---

* It is only while writing this work that I have become acquainted with the remarkable treatise of M. Lobatto, entitled *Théorie des Caracteristiques*, published at Amsterdam in 1837. It contains the theorem in the text, the analogous theorem in differential equations, and in one word the whole of the theory of linear equations with constant coefficients rediscovered in England a year or two afterwards, and published in the first and second volumes of the *Cambridge Mathematical Journal*. Every English mathematician will rejoice to see justice done to M. Lobatto.

It is proper to add that M. Lobatto's treatise does not contain any anticipation of the higher symbolical methods subsequently developed in this country.

$$(D-a)^{-i}a^x X = (\epsilon^{\frac{d}{dx}} - a)^{-i}\epsilon^{x\log a}X$$

$$= \epsilon^{x\log a}(\epsilon^{\frac{d}{dx}+\log a} - a)^{-i}X, \text{ by (7)},$$

$$= a^x(aD-a)^{-i}X$$

$$= a^{x-i}(D-1)^{-i}X$$

$$= a^{x-i}\Sigma^i X,$$

since $(D-1)^{-1} = \Delta^{-1} = \Sigma$. Hence, replacing $a^x X$ by X, and therefore X by $a^{-x}X$, we have

$$(D-a)^{-i}X = a^{x-i}\Sigma^i a^{-x}X \dots\dots\dots(8).$$

It is to be noted that since

$$\Sigma^i 0 = c_0 + c_1 x + c_2 x^2 \dots + c_{i-1}x^{i-1},$$

the above theorem, with the complementary function in its second member, will take the form

$$(D-a)^{-i}X = a^{x-i}\Sigma^i a^{-x}X + a^x(c_0 + c_1 x \dots + c_{i-1}x^{i-1})\dots(9).$$

If the equation $f(m)=0$ have a pair of imaginary roots $\alpha \pm \beta\sqrt{(-1)}$, that pair occurring only once, then, as appears from (5), the value of u will contain a pair of conjugate terms of the form

$$\{M+N\sqrt{(-1)}\}\{\alpha+\beta\sqrt{(-1)}\}^{x-1}\Sigma[\{\alpha+\beta\sqrt{(-1)}\}^{-x}X]$$

$$+\{M-N\sqrt{(-1)}\}\{\alpha-\beta\sqrt{(-1)}\}^{x-1}\Sigma[\{\alpha-\beta\sqrt{(-1)}\}^{-x}X].$$

Now let $\alpha = \rho\cos\theta$, $\beta = \rho\sin\theta$, whence $\theta = \tan^{-1}\frac{\beta}{\alpha}$, then

$$\{\alpha\pm\beta\sqrt{(-1)}\}^{x-1} = \rho^{x-1}\{\cos(x-1)\theta \pm \sqrt{(-1)}\sin(x-1)\theta\}$$

$$\{\alpha\pm\beta\sqrt{(-1)}\}^{-x} = \rho^{-x}\{\cos x\theta \mp \sqrt{(-1)}\sin x\theta\}.$$

Whence, substituting and reducing, we obtain as the real expression of the portion of the value of u corresponding to the imaginary roots in question,

$$2\rho^{x-1}\left[\begin{array}{l}\{M\cos(x-1)\theta - N\sin(x-1)\theta\}\Sigma\rho^{-x}X\cos(x\theta)\\+\{M\sin(x-1)\theta + N\cos(x-1)\theta\}\Sigma\rho^{-x}X\sin(x\theta)\end{array}\right]\dots(10).$$

The complementary function is evidently

$$2\rho^{x-1} \begin{bmatrix} \{M\cos(x-1)\,\theta - N\sin(x-1)\,\theta\}\,C_1 \\ +\{M\sin(x-1)\,\theta + N\cos(x-1)\,\theta\}\,C_2 \end{bmatrix},$$

which is reducible to the form

$$\rho^x\,(A\cos x\theta + B\sin x\theta).$$

If the imaginary pair of roots be repeated r times, then, as is shewn by (6), there will exist in u a series of pairs of conjugate terms of the form

$$\{M+N\sqrt{(-1)}\}\,\{\alpha+\beta\sqrt{(-1)}\}^{x-i}\Sigma^i\,[\{\alpha+\beta\sqrt{(-1)}\}^{-x}X]$$
$$+\{M-N\sqrt{(-1)}\}\,\{\alpha-\beta\sqrt{(-1)}\}^{x-i}\Sigma^i\,[\{\alpha-\beta\sqrt{(-1)}\}^{-x}X],$$

i receiving every integer value from 1 to r inclusive, and M and N being different for each different value of i. Proceeding as before, we find that the real expression of that portion of the value of u which corresponds to the imaginary roots in question will consist of a series of terms of the form

$$2\rho^{x-i}\begin{bmatrix} \{M\cos(x-i)\,\theta - N\sin(x-i)\,\theta\}\,\Sigma^i\rho^{-x}\cos x\theta X \\ +\{M\sin(x-i)\,\theta + N\cos(x-i)\,\theta\}\,\Sigma\,\rho^{-x}\sin x\theta X \end{bmatrix}..(11),$$

i varying from 1 to r.

It is evident also that the complementary function introduced by summation will ultimately be

$$\rho^x\begin{Bmatrix} (A_1+A_2x\dots+A_rx^{r-1})\cos x\theta \\ +(B_1+B_2x\dots+B_rx^{r-1})\sin x\theta \end{Bmatrix}\dots\dots\dots(12).$$

A_1, B_1, &c. being arbitrary constants.

7.   When the second member X either consists of a series of exponentials of the form ha^x, or is a rational and integral function of x, or is composed of terms resolvable into factors of either of these species, the process of solution may be simplified by methods analogous to those employed in the corresponding cases of differential equations; and, though it will still be necessary to determine the roots of the auxiliary equation $f(m)=0$, it will not be necessary to decompose the symbolic expression $\{f(D)\}^{-1}$.

In each of these cases, representing the given equation in the form $f(D) u = X$, we are permitted to write

$$u = \{f(D)\}^{-1} X + \{f(D)\}^{-1} 0 \dots\dots\dots(13),$$

the second term in the right-hand member representing the complementary function. And, as this function introduces the requisite number of constants, it suffices to deduce any particular value of the first term.

First, let X consist of a series of exponentials of the form above described. Then since

$$F(D) a^x = F(\epsilon^{\frac{d}{dx}}) \epsilon^{x \log a}$$
$$= F(\epsilon^{\log a}) \epsilon^{x \log a}$$
$$= F(a) a^x \dots\dots(14),$$

a particular value of $\{f(D)\}^{-1} X$ in (13) may be determined.

Cases of failure from $F(a)$ becoming infinite may be treated by combining with the term (11) such a term derived from the complementary function as will replace the infinite by an indeterminate form, and then applying the rule for vanishing fractions.

Secondly, let X be a rational and integral function of x. Then we may either convert D into $1 + \Delta$, and so develope $\{f(D)\}^{-1}$ in ascending powers of Δ, and finally perform the resulting operations, whether of summation or of difference, on X; or, converting D into $\epsilon^{\frac{d}{dx}}$, we may develope $\{f(D)\}^{-1}$ in ascending powers of $\frac{d}{dx}$, and perform on X the resulting operations of integration or differentiation. The solution is necessarily *finite*, since if X be of the nth degree both differences and differentials of an order higher than the nth vanish.

Thirdly, let X consist of terms of the form $a^x x^n$. Now

$$F(D) a^x \phi(x) = F(\epsilon^{\frac{d}{dx}}) \epsilon^{x \log a} \phi(x)$$
$$= \epsilon^{x \log a} F(\epsilon^{\frac{d}{dx} + \log a}) \phi(x)$$
$$= a^x F(aD) \phi(x).$$

Hence
$$F(D)\, a^x x^n = a^x F(aD)\, x^n.$$

By this theorem the problem is reduced to the second case already considered.

To the above cases various others arising from the introduction of circular functions may be reduced; but it is unnecessary to pursue the subject.

Lastly, it may be observed that when, in the application of the general method, the final summations can in no other way be effected, we must revert to the definition of the symbol Σ as in Ex. 3, and write down the series of terms which it may happen to indicate.

8.   It will be more instructive to exemplify the principles above explained than merely to apply the results which have been arrived at.

Ex. 8. $u_{x+2} - 5u_{x+1} + 6u_x = a^x.$

Here $u_x = (D^2 - 5D + 6)^{-1} a^x$

$$= \{(D-3)^{-1} - (D-2)^{-1}\}\, a^x$$

$$= 3^{x-1} \Sigma\, 3^{-x} a^x - 2^{x-1} \Sigma\, 2^{-x} a^x$$

$$= 3^{x-1} \Sigma \left(\frac{a}{3}\right)^x - 2^{x-1} \Sigma \left(\frac{a}{2}\right)^x$$

$$= 3^{x-1} \left\{ \frac{\left(\dfrac{a}{3}\right)^x}{\dfrac{a}{3} - 1} + C_1 \right\} - 2^{x-1} \left\{ \frac{\left(\dfrac{a}{2}\right)^x}{\dfrac{a}{2} - 1} + C_2 \right\}$$

$$= \frac{a^x}{a - 3} - \frac{a^x}{a - 2} + C3^x + C'2^x,$$

the complete primitive. Or, availing ourselves of the simplifications of Art. 7, we should proceed thus:

$$u_x = (D^2 - 5D + 6)^{-1} a^x + (D^2 - 5D + 6)^{-1} 0$$

$$= \frac{a^x}{a^2 - 5a + 6} + C3^x + C'2^x.$$

If $a = 3$ or 2, this solution fails. Suppose $a = 3$, then we may write

$$u_x = \text{limit of } \frac{a^x - 3^x}{a^2 - 5a + 6} + C3^x + C'2^x$$

$$= \frac{xa^{x-1}}{2a - 5} + C3^x + C'2^x$$

$$= x \times 3^{x-1} + C3^x + C'2^x.$$

Ex. 9.   Given

$$u_{x+2} - 4u_{x+1} + 4u_x = xa^x.$$

Symbolically we have

$$(D^2 - 4D + 4)\, u_x = xa^x;$$
$$\therefore\ u_x = (D - 2)^{-2} xa^x + (D - 2)^{-2} 0$$
$$= a^x (aD - 2)^{-2} x + 2^x (c_1 + c_2 x), \text{ by (11)}.$$

Now

$$(aD - 2)^{-2} x = (a - 2 + a\Delta)^{-2} x$$
$$= \{(a - 2)^{-2} - 2(a - 2)^{-3} a\Delta + \&c.\}\, x$$
$$= \frac{x}{(a - 2)^2} - \frac{2a}{(a - 2)^3}.$$

Hence, finally,

$$u_x = a^x \left\{ \frac{x}{(a - 2)^2} - \frac{2a}{(a - 2)^3} \right\} + 2^x (c + c'x).$$

Ex. 10. $u_{x+2} + a^2 u_x = \cos mx.$

Symbolically we have

$$(D^2 + a^2)\, u_x = \cos mx;$$
$$\therefore\ u_x = (D^2 + a^2)^{-1} \cos mx + (D^2 + a^2)^{-1} 0.$$

But $(D^2 + a^2)^{-1} \cos mx$

$$= \frac{1}{2} (D^2 + a^2)^{-1} \{\epsilon^{mx\sqrt{(-1)}} + \epsilon^{-mx\sqrt{(-1)}}\}$$

$$= \frac{1}{2} \left\{ \left(\epsilon^{2m\sqrt{(-1)}} + a^2 \right)^{-1} \epsilon^{mx\sqrt{(-1)}} + \left(\epsilon^{-2m\sqrt{(-1)}} + a^2 \right)^{-1} \epsilon^{-mx\sqrt{(-1)}} \right\}$$

$$= \frac{a^2 \cos mx + \cos m\,(x-2)}{a^4 + 2a^2 \cos 2m + 1} \quad \text{on reduction;}$$

and, by Ex. 6,

$$(D^2 + a^2)^{-1}\,0 = a^x \left(A \cos \frac{\pi x}{2} + B \sin \frac{\pi x}{2} \right);$$

$$\therefore\; u_x = \frac{a^2 \cos mx + \cos m\,(x-2)}{a^4 + 2a^2 \cos 2m + 1} + a^x \left(A \cos \frac{\pi x}{2} + B \sin \frac{\pi x}{2} \right).$$

Ex. 11. $u_{x+2} - a^2 u_x = \tan mx.$

Proceeding by the general method we have

$$u_x = (D^2 - a^2)^{-1} \tan mx + (D^2 - a^2)^{-1}\,0$$

$$= \frac{1}{2a} \left\{ (D - a)^{-1} - (D + a)^{-1} \right\} \tan mx + C a^x + C'\,(-a)^x$$

$$= \frac{1}{2a} \left\{ a^{x-1} \Sigma a^{-x} \tan mx - (-a)^{x-1} \Sigma\,(-a)^{-x} \tan mx \right\}$$
$$+ C a^x + C'\,(-a)^x.$$

As the summation cannot be effected in definite terms we may write

$$\Sigma a^{-x} \tan mx = \left\{ \frac{\tan m}{a} + \frac{\tan 2m}{a^2} \cdots + \frac{\tan\,(x-1)\,m}{a^{x-1}} \right\},$$

with a corresponding form for $\Sigma\,(-a)^{-x} \tan mx.$

Equations reducible to linear equations with constant coefficients.

9.   There are certain forms of equations which a transformation enables us to reduce to the class of linear equations with constant coefficients above considered. Far less attention has indeed been paid to such reductions than to the corresponding ones in differential equations, and the number of known cases is small. The following are the most important.

1st. Equations of the form

$$u_{x+n} + A_1 \phi(x) u_{x+n-1} + A_2 \phi(x) \phi(x-1) u_{x+n-2}$$
$$+ A_3 \phi(x) \phi(x-1) \phi(x-2) u_{x+n-3} + \&c. = X \ldots\ldots (1),$$

where $A_1 A_2 \ldots A_n$ are constant, and $\phi(x)$ a known function, may be reduced to equations with constant coefficients by assuming

$$u_x = \phi(x-n) \phi(x-n-1) \ldots \phi(1) v_x \ldots\ldots\ldots\ldots (2).$$

For this substitution gives

$$u_{x+n} = \phi(x) \phi(x-1) \phi(x-2) \ldots \phi(1) v_{x+n},$$

$$u_{x+n-1} = \phi(x-1) \phi(x-2) \ldots \phi(1) v_{x+n-1},$$

and so on; whence substituting and dividing by the common factor $\phi(x) \phi(x-1) \ldots \phi(1)$, we get,

$$v_{x+n} + A_1 v_{x+n-1} + A_2 v_{x+n-2} + \&c. = \frac{X}{\phi(x) \phi(x-1) \ldots \phi(1)}$$
$$\ldots\ldots\ldots (3),$$

an equation with constant coefficients.

In effecting the above transformation we have supposed x to admit of a system of positive integral values. The general transformation would obviously be

$$u_x = \phi(x-n) \phi(x-n-1) \ldots \phi(r),$$

r being any particular value of x assumed as *initial*.

2ndly. Equations of the form

$$u_{x+n} + A_1 a^x u_{x+n-1} + A_2 a^{2x} u_{x+n-2} + \&c. = X,$$

are virtually included in the above class. For, assuming $\phi(x) = a^x$, they may be presented in the form

$$u_{x+n} + A_1 \phi(x) u_{x+n-1} + \frac{A_2}{a} \phi(x) \phi(x-1) u_{x+n-2} + \&c. = X$$
$$\ldots\ldots\ldots (4).$$

Hence, to integrate them it is only necessary to assume

$$u_x = a^{\{1+2+3\ldots+(x-n)\}} v_x$$

$$= a^{\dfrac{(x-n)(x-n+1)}{2}} v_x \ldots\ldots\ldots\ldots\ldots (5).$$

3rdly. Equations of the form

$$u_{x+1} u_x + a_x u_{x+1} + b_x u_x = c_x \ldots\ldots\ldots\ldots (6)$$

can be reduced to linear equations of the second order, and, under certain conditions, to linear equations with constant coefficients.

Assume

$$u_x = \frac{v_{x+1}}{v_x} - a_x.$$

Then for the first two terms of the proposed equation, we have

$$u_{x+1} (u_x + a_x) = \left(\frac{v_{x+2}}{v_{x+1}} - a_{x+1} \right) \frac{v_{x+1}}{v_x}$$

$$= \frac{v_{x+2}}{v_x} - a_{x+1} \frac{v_{x+1}}{v_x}.$$

Whence substituting and reducing, we find

$$v_{x+2} + (b_x - a_{x+1}) v_{x+1} - (a_x b_x + c_x) v_x = 0 \ldots\ldots\ldots (7),$$

a linear equation whose coefficients will be constant if the functions $b_x - a_{x+1}$ and $a_x b_x + c_x$ are constant, and which again by the previous section may be reduced to an equation with constant coefficients if those functions are of the respective forms

$$A\phi(x), \quad B\phi(x)\phi(x-1).$$

The above equation may also be integrated when the single condition $c_x = 0$ is satisfied. For, on separating the symbols, (7) becomes

$$\{D^2 + (b_x - a_{x+1}) D - a_x b_x\} v_x = 0.$$

Now this may be reduced to the form

$$(D + b_x)(D - a_x) v_x = 0,$$

which, lastly, may be resolved into the two linear equations of the first order,

$$\left. \begin{array}{l} (D - a_x)\, v_x = w_x \\ (D + b_x)\, w_x = 0 \end{array} \right\} \quad \dots\dots\dots\dots\dots\dots (8).$$

We can now determine in succession w_x, v_x and u_x.

As the value of v_x is in all these cases made to depend upon the solution of an equation of differences of the second order, it will involve two arbitrary constants, but they will effectively be reduced to one in the final substitution of the derived value of u_x in (6).

4thly. Some non-linear equations may be solved by means of the relations which connect the successive values of circular functions.

Ex. 12. $u_{x+1} u_x - a_x \left(u_{x+1} - u_x \right) + 1 = 0.$

Here we have

$$\frac{1}{a_x} = \frac{u_{x+1} - u_x}{1 + u_{x+1} u_x}.$$

Now the form of the second member suggests the transformation $u_x = \tan v_x$, which gives

$$\frac{1}{a_x} = \frac{\tan v_{x+1} - \tan v_x}{1 + \tan v_{x+1} \tan v_x}$$

$$= \tan \left(v_{x+1} - v_x \right)$$

$$= \tan \Delta v_x,$$

whence

$$v_x = C + \Sigma \tan^{-1} \frac{1}{a_x},$$

$$u_x = \tan \left(C + \Sigma \tan^{-1} \frac{1}{a_x} \right).$$

Ex. 13.   Given $u_{x+1} u_x + \sqrt{\{(1 - u^2_{x+1})\,(1 - u_x{}^2)\}} = a_x.$

Let $u_x = \cos v_x$, and we have

$$a_x = \cos v_{x+1} \cos v_x + \sin v_{x+1} \sin v_x$$

$$= \cos \left(v_{x+1} - v_x \right) = \cos \Delta v_x,$$

whence finally

$$u_x = \cos \left(C + \Sigma \cos^{-1} a_x \right).$$

The method by which the linear equation (7) was integrated in the particular case of $c_x = 0$ suggests the problem of determining *a priori* the forms of equations of differences of the higher orders which may be resolved into, or in some way made to depend upon, equations of the first order. The following will serve as a particular illustration.

Since the solution of the symbolical equation

$$(\pi^2 + a\pi + b) \, u_x = X \, \dotfill (9),$$

where a and b are constant coefficients, depends in general upon the interpretation of an expression of the form

$$u_x = N_1 \, (\pi - \alpha)^{-1} X + N_2 \, (\pi - \beta)^{-1} X \dotfill (10),$$

let us give to π the form $D + \phi \, (x)$; such being a form which renders (10) interpretable by the solution of linear equations of differences of the first order. Then (9) becomes

$$[\{D + \phi \, (x)\}^2 + a \, \{D + \phi \, (x)\} + b] \, u_x = X,$$

or, effecting the operations and replacing Du_x, $D^2 u_x$ by u_{x+1}, u_{x+2},

$$u_{x+2} + \{\phi \, (x) + \phi \, (x+1) + a\} u_{x+1} + [\{\phi(x)\}^2 + a\phi \, (x) + b] \, u_x = X.$$

The solution of this equation then, whatever form we assign to ϕ, will depend upon that of equations of differences of the first order.

There are, further, various cases in which the solution of equations of differences may be effected by a process of successive reduction. We shall discuss this subject in Chap. IX.

Analogy with Differential Equations.

10. Many remarkable properties which constitute a ground of analogy more or less exact between linear differential equations and algebraic equations, may be extended to linear equations of differences.

One of those properties which, though among the least interesting in point of analogy, is the most important in application, we shall here notice.

THEOREM. *We can depress by unity the order of a linear equation of differences*

$$u_{x+n} + A_x u_{x+n-1} + B_x u_{x+n-2} + \&c. = X \dots\dots (1),$$

if we know a particular value of u_x which would satisfy it were the second member 0.

Let v_x be such a value, so that

$$v_{x+n} + A_x v_{x+n-1} + B_x v_{x+n-2} + \&c. = 0 \dots\dots (2),$$

and let $u_x = v_x t_x$; then (1) becomes

$$v_{x+n} t_{x+n} + A_x v_{x+n-1} t_{x+n-1} + B_x v_{x+n-2} t_{x+n-2} + \&c. = X.$$

Or $v_{x+n} D^n t_x + A_x v_{x+n-1} D^{n-1} t_x + B_x v_{x+n-2} D^{n-2} t_x \dots = X.$

Replacing D by $1 + \Delta$, and developing D^n, D^{n-1}, &c. in ascending powers of Δ, arrange the result according to ascending differences of t_x. There will ensue

$$(v_{x+n} + A_x v_{x+n-1} + B_x v_{x+n-2} \dots) t_x$$
$$+ P\Delta t_x + Q\Delta^2 t_x \dots + Z\Delta^n t_x = X \dots\dots (3),$$

P, Q, ... Z being, like the coefficient of t_x, functions of v_x, v_{x+1}, &c. and of the original coefficients A_x, B_x, &c.

Now the coefficient of t_x vanishes by (2), whence, making $\Delta t_x = w_x$, we have

$$P w_x + Q\Delta w_x \dots + Z\Delta^{n-1} w_x = X,$$

an equation of differences of the $n - 1^{\text{th}}$ order for determining w_x. This being found we have

$$t_x = \Sigma w_x; \quad \therefore \ u_x = v_x \Sigma w_x \dots\dots (4).$$

It hence follows that the linear equation

$$u_{x+2} + A_x u_{x+1} + B_x u_x = X \dots\dots\dots\dots(5)$$

can be completely integrated if we know a particular integral of

$$u_{x+2} + A_x u_{x+1} + B_x u_x = 0 \dots\dots\dots\dots(6).$$

To deduce the form of the complete integral, let, as before, v_x be the particular value of u_x which satisfies (6), and let $u_x = v_x t_x$; then, proceeding as above, we get

$$v_{x+2} \Delta^2 t_x + (2v_{x+2} + A_x v_{x+1}) \Delta t_x = X;$$

therefore putting $\Delta t_x = w_x$ and dividing by v_{x+2},

$$v_{x+2} \Delta w_x + (2v_{x+2} + A_x v_{x+1}) w_x = X,$$

$$\Delta w_x + \left(2 + A_x \frac{v_{x+1}}{v_{x+2}}\right) w_x = \frac{X}{v_{x+2}};$$

or
$$w_{x+1} + \left(1 + A_x \frac{v_{x+1}}{v_{x+2}}\right) w_x = \frac{X}{v_{x+2}}.$$

This equation may be a little simplified; for by virtue of (6), of which v_x is an integral,

$$1 + \frac{A_x v_{x+1}}{v_{x+2}} = -\frac{B_x v_x}{v_{x+2}}.$$

Hence

$$w_{x+1} - \frac{B_x v_x}{v_{x+2}} w_x = \frac{X}{v_{x+2}}.$$

Ultimately therefore we have

$$u_x = v_x \Sigma \left\{ \left(P \frac{B_{x-1} v_{x-1}}{v_{x+1}} \right) \Sigma \frac{X}{v_{x+2} P \dfrac{B_x v_x}{v_{x+2}}} \right\},$$

each summation introducing an arbitrary constant.

If we make

$$P \frac{B_{x-1} v_{x-1}}{v_{x+1}} = T_x,$$

the above solution may be presented in the form

$$u_x = v_x \Sigma \left(T_x \Sigma \frac{X}{B_x v_x T_x} \right).$$

If $X = 0$ we have simply

$$u_x = v_x \left(C + C_1 \Sigma T_x \right).$$

Fundamental Connexion with Differential Equations.

11. Equations of differences are connected with differential equations by more than mere analogy. If in an equation of differences symbolically expressed we substitute $\epsilon^{\frac{d}{dx}}$ for D, it becomes a differential equation. We purpose here to inquire to what form of the general solution of a linear equation of differences, with constant coefficients and with vanishing second member, this transformation leads.

This solution must be deduced from the interpretation of terms of the form

$$M_r (\epsilon^{\frac{d}{dx}} - a)^{-r} 0 \dots\dots\dots\dots (1),$$

a, r, and M having the same meaning and the same value as in Art. 6. We know from the theory of differential equations that the complete interpretation of this term will consist of terms of the form

$$\epsilon^{mx} (A + Bx + Cx^2 \dots + Ex^{r-1}) \dots\dots\dots (2),$$

m representing in succession the different roots of the equation

$$\epsilon^m - a = 0 \dots\dots\dots\dots\dots (3),$$

and A, B, C, ... E being arbitrary constants which differ for different values of m.

Now the roots of (3) are all included in the formula

$$m = \log a \pm 2i\pi \sqrt{(-1)},$$

i being an integer, positive, negative, or 0.

Let $i = 0$, then $m = \log a$, and (2) gives

$$a^x (A + Bx + Cx^2 \ldots + Ex^{r-1}),$$

which agrees with the result in Art. 6.

Again, assigning to i any two corresponding positive and negative values, (2) assumes the form

$$S e^{\{\log a + 2i\pi \sqrt{(-1)}\}x} + T e^{\{\log a - 2i\pi \sqrt{(-1)}\}x} \ldots \ldots \ldots \ldots (4),$$

S and T being polynomials of the form

$$A + Bx + Cx^2 \ldots + Ex^{r-1},$$

but with different constants. But (4) may be reduced to the form

$$a^x \left[S \{\cos 2i\pi x + \sqrt{(-1)} \sin 2i\pi x\} \right.$$
$$\left. + T \{\cos 2i\pi x - \sqrt{(-1)} \sin 2i\pi x\} \right],$$

or, replacing $S + T$ and $(S - T)\sqrt{(-1)}$, by M and N, (polynomials still of the $r - 1^{\text{th}}$ degree with arbitrary constant coefficients), we have

$$a^x (M \cos 2i\pi x + N \sin 2i\pi x).$$

Hence, giving to i the successive values 0, 1, 2, &c., it is seen that the *complete* solution, so far as it depends on the expression $(D - a)^{-r} 0$, will be of the form

$$a^x (P_1 + P_2 x + P_3 x^2 \ldots + P_r x^{r-1}) \ldots \ldots \ldots \ldots (5),$$

where each coefficient P is of the form

$$A + \begin{cases} B_1 \cos 2\pi x + C_1 \cos 4\pi x + E_1 \cos 6\pi x + \&c. \\ B_2 \sin 2\pi x + C_2 \sin 4\pi x + E_2 \sin 6\pi x + \&c. \end{cases} \ldots (6),$$

A, B_1, B_2, &c., being arbitrary constants.

Analytically, (6) is the general expression for a periodical function of x constant for values of x differing by unity. And this confirms the *à priori* determination of the significance of the constants in the solution of equations of differences. Art. 3. See also Chap. IV. Art. 1.

We thus see that it is due to the imaginary constituents of the expression $e^{\frac{d}{dx}} - a$ that the constants in the solution of an equation of differences are such only relatively and not absolutely, their true character being periodical.

EXERCISES.

1. Find the equations of differences to which the following complete primitives belong.

1st. $u = cx^2 + c^2$.   2nd. $u = \left\{ c\,(-1)^x - \dfrac{x}{2} \right\}^2 - \dfrac{x^4}{4}$.

3rd. $u = cx + c'a^x$.   4th. $u = ca^x + c^2$.

5th. $u = c^2 + c\left(\dfrac{1-a}{1+a}\right)(-a)^x - \dfrac{a^{2x+1}}{(1+a)^2}$.

2. Given $u_{x+2} - 3u_{x+1} - 4u_x = 0$, find u_x.

3. $u_{x+2} - 3u_{x+1} - 4u_x = m^x$.

4. $u_{x+1} - pa^{2x}u_x = qa^{x^2}$.

5. $u_{x+1} - au_x = \cos nx$.

6. $u_{x+2} + 4u_{x+1} + 4u_x = x$.

7. $u_x - 3\sin(\pi x)u_{x-1} - 2\{\sin(\pi x)\}^2 u_{x-2} = 0$.

8. The linear equation of differences $u_x - A_x u_{x-1} = B_x$, becomes integrable on multiplying both sides by PA_x.

9. Deduce a complete primitive of the equation
$$u_{x+2} + 2u_{x+1} + u_x = x\,(x-1)\,(x-2).$$

10. Integrate $u_{x+2} - 2mu_{x+1} + (m^2 + n^2)u_x = 0$.

11. Integrate $u_{x+1}u_x + (x+2)u_{x+1} + xu_x = -2 - 2x - x^2$.

12. In Art. 9, the solution of a class of equations of the

first order is made to depend upon that of a linear equation of the second order whose second member is 0 by assuming

$$u_x = \frac{v_{x+1}}{v_x} - a_x.$$

And it is remarked that the two constants which appear in the value of v_x effectively produce only one in that of u_x. Prove this.

13. The equation

$$u_{x+2} - (a^{x+1} + a^{-x}) u_{x+1} + u_x = 0$$

may be resolved into two equations of differences of the first order.

14. Given that a particular solution of the equation

$$u_{x+2} - a(a^x + 1) u_{x+1} + a^{x+1} u_x = 0 \text{ is } u_x = ca^{\frac{x(x-1)}{2}},$$

deduce the general solution.

15. The above equation may be solved without the previous knowledge of a particular integral.

16. The equation

$$u_x u_{x+1} u_{x+2} = a(u_x + u_{x+1} + u_{x+2})$$

may be integrated by assuming $u_x = \sqrt{a} \tan v_x$.

17. Shew also that the general integral of the above equation is included in that of the equation $u_{x+3} - u_x = 0$, and hence deduce the former.

18. Shew how to integrate the equation

$$u_{x+1} u_{x+2} + u_{x+2} u_x + u_x u_{x+1} = m^2.$$

CHAPTER VIII.

OF EQUATIONS OF DIFFERENCES OF THE FIRST ORDER, BUT NOT OF THE FIRST DEGREE.

1. THE theory of equations of differences which are of a degree higher than the first differs much from that of the corresponding class of differential equations, but it throws upon the latter so remarkable a light that for this end alone it would be deserving of attentive study. We shall endeavour to keep the connexion of the two subjects in view throughout this chapter.

Expressing an equation of differences of the first order and n^{th} degree in the form

$$(\Delta u)^n + P_1(\Delta u)^{n-1} + P_2(\Delta u)^{n-2} \dots + P_n u = Q \dots\dots\dots (1),$$

$P_1 P_2 \dots P_n$ and Q being functions of the variables x and u, and then by algebraic solution reducing it to the form

$$(\Delta u - p_1)(\Delta u - p_2) \dots (\Delta u - p_n) = 0 \dots\dots\dots(2),$$

it is evident that the complete primitive of any one of the component equations,

$$\Delta u - p_1 = 0, \quad \Delta u - p_2 = 0 \dots \Delta u - p_n = 0 \dots\dots (3),$$

will be a complete primitive of the given equation (1), i.e. *a solution involving an arbitrary constant*. And thus far there is complete analogy with differential equations (*Diff. Equations*, Chap. VII. Art. 1). But here a first point of difference arises. The complete primitives of a differential equation of the first order, obtained by resolution of the equation with respect to $\frac{dy}{dx}$ and solution of the component equations, may without loss of generality be replaced by a single complete primitive. (*1b.* Art. 3). Referring to the demonstration of this, the reader will see that it depends mainly upon the fact that the differential coefficient with respect to x of any function of V_1, V_2, ... V_n, variables supposed dependent on x, will be linear with respect to the differential coefficients of these dependent variables (*Ib.* (16) (17)). But this property does not remain if the operation Δ is substituted for that of $\dfrac{d}{dx}$; and therefore the different complete primitives of an equation of differences cannot be replaced by a single complete primitive. On the contrary, it may be shewn that out of the complete primitives corresponding to the component equations into which the given equation of differences is supposed to be resolvable, an infinite number of other complete primitives may be evolved corresponding, not to particular component equations, but to a *system* of such components succeeding each other according to a determinate law of alternation as the independent variable x passes through its successive values.

Ex. 1. Thus suppose the given equation to be

$$(\Delta u)^2 - (a+x)\,\Delta u + ax = 0 \,\ldots\ldots\ldots\ldots (4),$$

which is resolvable into the two equations

$$\Delta u - a = 0, \quad \Delta u - x = 0 \,\ldots\ldots\ldots\ldots\ldots(5).$$

And suppose it required to obtain a complete primitive which shall satisfy the given equation (4) by satisfying the first of the component equations (5) when x is an even integer and the second when x is an odd integer.

The condition that Δu shall be equal to a when x is even, and to x when x is odd, is satisfied if we assume

$$\Delta u = a\,\frac{1 + (-1)^x}{2} + x\,\frac{1 - (-1)^x}{2}$$

$$= \frac{a+x}{2} + (-1)^x\,\frac{a-x}{2}\,,$$

the solution of which is

$$u = \frac{ax}{2} + \frac{x\,(x-1)}{4} + (-1)^x\left(\frac{x-a}{4} - \frac{1}{8}\right) + C,$$

and it will be found that this value of u satisfies the given equation in the manner prescribed. Moreover, it is a complete primitive.

To extend this method of solution to any proposed equation and to any proposed case, it is only necessary to express Δu as a linear function of the particular values which it is intended that it should receive, each such value being multiplied by a coefficient which has the property of becoming equal to unity for the values of x for which that term becomes the equivalent of Δu, and to 0 for all other values. The forms of the coefficients may be determined by the following proposition.

PROP. If α, β, γ ... be the several n^{th} roots of unity, then x being an integer, the function $\dfrac{\alpha^x + \beta^x + \gamma^x \cdots}{n}$ is equal to unity if x be equal to n or a multiple of n, and is equal to 0 if x be not a multiple of n.

For if $\mu = \cos \dfrac{2\pi}{n} + \sqrt{(-1)} \sin \dfrac{2\pi}{n}$, the n roots will be $1, \mu, \mu^2 \ldots \mu^{n-1}$. Therefore,

$$\frac{\alpha^x + \beta^x + \gamma^x \cdots}{n} = \frac{1 + \mu^x + \mu^{2x} \ldots + \mu^{(n-1)x}}{n}$$

$$= \frac{1}{n} \frac{\mu^{nx} - 1}{\mu^x - 1} \ldots \ldots \ldots \ldots \ldots \ldots (6).$$

Now if x be equal to n, or to a multiple of n, the above becomes a vanishing fraction whose value determined in the usual way is unity. If x be not a multiple of n, then since $\mu^n = 1$, the numerator vanishes while the denominator does not, and the fraction is therefore equal to 0.

Hence, if it be required to form such an expression for Δu as shall assume the particular values $p_1, p_2 \ldots p_n$ in succession for the values $x = 1$, $x = 2 \ldots x = n$, again, for the values $x = n + 1$, $x = n + 2 \ldots x = 2n$, and so on, *ad inf.*, it suffices to assume

$$\Delta u = P_{x-1} p_1 + P_{x-2} p_2 \cdots + P_{x-n} p_n \ldots \ldots \ldots \ldots (7),$$

where $$P_x = \frac{\alpha^x + \beta^x + \gamma^x \cdots}{n} \ldots \ldots \ldots \ldots \ldots (8),$$

α, β, γ ... being as above the different n^{th} roots of unity. The equation (7) must then be integrated.

It will be observed that the same values of Δu may recur in any order. Further illustration than is afforded by Ex. 1, is not needed. Indeed, what is of chief importance to be noted is not the method of solution, which might be varied, but the nature of the connexion of the derived complete primitives with the complete primitives of the component equations into which the given equation of differences is resolvable. It is seen that any one of those derived primitives would geometrically form a sort of connecting envelope of the loci of what may be termed its *component* primitives, i. e. the complete primitives of the component equations of the given equation of differences.

If x be the abscissa, u_x the corresponding ordinate of a point on a plane referred to rectangular axes, then any particular primitive of an equation of differences represents a system of such points, and a complete primitive represents an infinite number of such systems. Now let two consecutive points in any system be said to constitute an element of that system, then it is seen that the successive elements of any one of these systems of points representing the locus of a *derived* primitive (according to the definitions implied above) will be taken in a determinate cyclical order from the elements of systems corresponding to what we have termed its *component* primitives.

2. It is possible also to deduce new complete primitives from a single complete primitive, provided that in the latter the expression for u be of a higher degree than the first with respect to the arbitrary constant. The method which consists in treating the constant as a variable parameter, and which leads to results of great interest from their connexion with the theory of differential equations, will be exemplified in the following section.

Solutions derived from the Variation of a Constant.

A given complete primitive of an equation of differences of the first order being expressed in the form

$$? \quad y = f(x, c) \quad \dots\dots\dots\dots\dots\dots(1),$$

let c vary, but under the condition that Δu shall admit of the same expression in terms of x and c as if c were a constant. It is evident that if the value of c determined by this condition as a function of x be substituted in the given primitive (9) we shall obtain a new solution of the given equation of differences. The process is *analogous* to that by which from the complete primitive of a differential equation we deduce the singular solution, but it differs as to the character of the result. The solutions at which we arrive are not singular solutions, but new complete primitives, the condition to which c is made subject leading us not, as in the case of differential equations, to an algebraic equation for its discovery, but to an equation of differences the solution of which introduces a new arbitrary constant.

The new complete primitive is usually termed an indirect integral.

Ex. 2.   The equation $u = x\Delta u + (\Delta u)^2$ has for a complete primitive

$$u = cx + c^2 \dots\dots\dots\dots\dots\dots (2).$$

An indirect integral is required.

Taking the difference on the hypothesis that c is constant, we have

$$\Delta u = c \;;$$

and taking the difference of (2) on the hypothesis that c is an unknown function of x, we have

$$\Delta u = c + (x + 1)\,\Delta c + 2c\Delta c + (\Delta c)^2.$$

Whence, equating these values of Δu, we have

$$\Delta c\,(x + 1 + 2c + \Delta c) = 0 \dots\dots\dots\dots (3).$$

Of the two component equations here implied, viz.

$$\Delta c = 0, \quad \Delta c + 2c + x + 1 = 0,$$

the first determines c as an arbitrary constant, and leads back to the given primitive (2); the second gives, on integration,

$$c = C\,(-1)^x - \frac{x}{2} - \frac{1}{4} \dots\dots\dots\dots\dots(4),$$

C being an arbitrary constant, and this value of c substituted in the complete primitive (2) gives on reduction

$$u = \left\{ C \, (-1)^x - \frac{1}{4} \right\}^2 - \frac{x^2}{4} \dotfill (5).$$

Now this is an indirect integral. We see that the *principle* on which its determination rests is that upon which rests the deduction of the singular solutions of differential equations from their complete primitives. But in *form* the result is itself a complete primitive; and the reader will easily verify that it satisfies the given equation of differences without any particular determination of the constant C.

Again, as by the method of Art. 1 we can deduce from (3) an infinite number of complete primitives determining c, we can, by the substitution of their values in (2) deduce an infinite number of indirect integrals of the equation of differences given.

It is proper to observe that indirect integrals may be deduced from the equation of differences (provided that we can effect the requisite integrations) without the prior knowledge of a complete primitive.

Ex. 3. Thus assuming the equation of differences,

$$u = x\Delta u + (\Delta u)^2 \dotfill (6),$$

and taking the difference of both sides, we have

$$\Delta u = \Delta u + x\Delta^2 u + \Delta^2 u + 2\Delta u \Delta^2 u + (\Delta^2 u)^2;$$

$$\therefore \; \Delta^2 u \, (\Delta^2 u + 2\Delta u + x + 1) = 0,$$

which is resolvable into

$$\Delta^2 u = 0, \quad \Delta^2 u + 2\Delta u + x + 1 = 0 \dotfill (7).$$

The former gives on integrating once,

$$\Delta u = c,$$

and leads on substitution in the given equation to the complete primitive (2).

The second equation of (7) gives, after one integration,

$$\Delta u = C \, (-1)^x - \frac{x}{2} - \frac{1}{4},$$

and substituting this in (6) we have on reduction

$$u = \left\{ C \left(-1\right)^x - \frac{1}{4} \right\}^2 - \frac{x^2}{4},$$

which agrees with (5).

The process by which from a given complete primitive we deduce an indirect integral admits of geometrical interpretation.

For each value of c the complete primitive $u = f(x, c)$ may be understood to represent a system of points situated in a plane and referred to rectangular co-ordinates; the changing of c into $c + \Delta c$ then represents a transition from one such system to another. If such change leave unchanged the values of u and of Δu corresponding to a particular value of x, it indicates that there are two consecutive points, i. e. an *element* (Art. 1) of the system represented by $u = f(x, c)$, the position of which the transition does not affect. And the successive change of c as a function of x ever satisfying this condition indicates that each system of points formed in succession has one element common with the system by which it was preceded, and the next element common with the system by which it is followed. The system of points formed of these consecutive common elements is the so-called *indirect integral*, which is thus seen to be a connecting envelope of the different systems of points represented by the given complete primitive. The difference between this case and the one considered before is that here all the elements of all possible indirect integrals are, virtually, contained in the one complete primitive given.

It evidently might happen that both methods admitted of application in the same problem.

3.   Of the different questions which the above theory suggests, the following are perhaps the most important.

1st.   An indirect integral being itself a complete primitive, what will be the result of applying to it the process of the variation of the arbitrary constant?

2ndly.   An indirect integral represents a system of the loci represented by the complete primitive from which it was

9—2

derived, and it is itself a complete primitive. In differential
equations envelopes are represented usually by singular solutions, and occasionally by particular primitives. How is this
to be explained?

The answers to these questions are contained in the statements of two laws, viz. the Law of Reciprocity, connecting
the integrals of an equation of differences; and the Law of
Continuity, governing the transition from the integrals of
equations of differences to the integrals of differential equations.

Law of Reciprocity.

PROP. A complete primitive of an equation of differences
with the indirect integrals deducible from that primitive by
the variation of its arbitrary constant will together constitute
a cycle of complete primitives, such that from any one of
them all the others may be deduced by the variation of its
arbitrary constant.

Let the given complete primitive be

$$u = f(x, c)\dots\dots\dots (1).$$

Then

$$\Delta u = f(x+1, c) - f(x, c) \dots\dots (2),$$

and the elimination of c between these equations leads to the
equation of differences, the form of which will be

$$F(x, u, \Delta u) = 0\dots\dots\dots (3).$$

Again, taking the difference of (1) on the assumption that c
is an unknown function of x, we have

$$\Delta u = f(x+1, c+\Delta c) - f(x, c).$$

That this may be equal to the value of Δu obtained in (2) on
the hypothesis of c being a constant, we must have

$$f(x+1, c+\Delta c) - f(x+1, c) = 0\dots\dots (4),$$

an equation of differences for determining c. It is satisfied
by the assumption $\Delta c = 0$; but this, determining c as a constant,
only leads us back to the given complete primitive. Virtually
(4) is reducible to the form

$$\Delta c \phi(x, c, \Delta c) = 0 \dots\dots\dots(5),$$

and therefore resolvable into

$$\Delta c = 0, \quad \phi\,(x, c, \Delta c) = 0 \dots\dots\dots (6),$$

and it is the values of c which satisfy the second of these equations, or which, according to the theory developed in Art. 1, satisfy the two equations under some law of alternation, that lead on substitution in (1) to indirect integrals.

If $D = 1 + \Delta$, the equation (4) may be expressed in the form

$$f\,(x+1, Dc) - f\,(x+1, c) = 0 \dots\dots\dots (7).$$

Let its integrals, in accordance with what has above been said, be

$$c = a,$$
$$c = \phi_1\,(x, a_1), \quad c = \phi_2\,(x, a_2), \text{ &c.} \Big\} \dots\dots\dots (8),$$

$a, a_1, a_2 \dots$ being arbitrary constants.

Then the indirect integrals derivable from (1) are

$$u = f\,\{x, \phi_1\,(x, a_1)\}, \quad u = f\,\{x, \phi_2\,(x, a_2)\}, \text{ &c.} \dots (9).$$

But, since $c = \phi_1\,(x, a_1)$ is by hypothesis an integral of (7), it must, together with the value of Dc which it gives, viz.

$$Dc = \phi_1\,(x+1, a_1),$$

reduce (7) to an identity. Substituting in (7) we have

$$f\,\{x+1, \phi_1\,(x+1, a_1)\} - f\,\{x+1, \phi_1\,(x, a_1)\} = 0 \dots (10).$$

Proceeding in like manner with the equation $c = \phi_2\,(x, a_2)$ we have

$$f\,\{x+1, \phi_2\,(x+1, a_2)\} - f\,\{x+1, \phi_2\,(x, a_2)\} = 0 \dots (11),$$

and so on. *Thus the indirect integrals of* (3) *are represented by the system* (9), *the functions denoted by* ϕ_1, ϕ_2, *&c., being subject to the conditions* (10), (11), *&c.*

Now let us assume the first indirect integral in (9), viz.

$$u = f\,\{x, \phi_1\,(x, a_1)\} \dots\dots\dots (12),$$

as a *given* complete primitive, and seek the indirect integrals.

The form of the equation (7) shews that the equation of differences for determining a_1 will be

$$f\{x+1, \phi_1(x+1, Da_1)\} - f\{x+1, \phi_1(x+1, a_1)\} = 0.$$

But the equation of condition (10) enables us to reduce this to the form

$$f\{x+1, \phi_1(x+1, Da_1)\} - f\{x+1, \phi_1(x, a_1)\} = 0,$$

or, since $D\phi_1(x, a_1) = \phi_1(x+1, Da_1)$,

$$f\{x+1, D\phi_1(x, a_1)\} - f\{x+1, \phi_1(x, a_1)\} = 0 \dots (13);$$

an equation which is of the same form with respect to the function $\phi_1(x, a_1)$ as the equation (7) is with respect to c. Hence by (8) it will admit of the system of integrals,

$$\left.\begin{array}{l} \phi_1(x, a_1) = A, \\ \phi_1(x, a_1) = \phi_1(x, A_1), \quad \phi_1(x, a_1) = \phi_2(x, A_2), \&c. \end{array}\right\} \dots (14),$$

$A, A_1, A_2 \dots$ being arbitrary constants. These integrals determine the values of a_1, which must be substituted in the complete primitive (12) in order to obtain the system of its derived primitives.

The integral in the first line of (14) reduces (12) to the form

$$u = f(x, A),$$

which is equivalent to the original complete primitive (1).

The first integral in the second line of (14) gives $a_1 = A_1$, and thus leads us back to the assumed primitive (12).

The second integral in the second line of (14) reduces (12) to the form

$$u = f\{x, \phi_2(x, A_2)\},$$

which agrees with the second indirect integral in the system (9). And the forms being general, every integral in the system (9) can thus be derived from the first of the series.

We see therefore that any indirect integral leads us, by a repetition of the process of its own formation, both to the complete primitive from which it was derived, and to all other complete primitives having the same origin. The entire system is seen to constitute a cycle. Theoretically, that cycle will be infinite, but practically, owing to the limitation of our powers of integration, it may be finite.

Ex. 4. The equation of differences, $u = x\Delta u + (\Delta u)^2$, has for a complete primitive $u = cx + c^2$, and hence (2, Ex. 2) was derived the indirect integral

$$u = \left\{ C(-1)^x - \frac{1}{4} \right\}^2 - \frac{x^2}{4} ;$$

shew that, assuming this as complete primitive, the equation $u = cx + c^2$ results as indirect integral.

Expressing the given primitive in the form

$$u = C^2(-1)^{2x} - \frac{C}{2}(-1)^x + \frac{1}{16} - \frac{x^2}{4} \quad \ldots\ldots\ldots\ldots(a),$$

and taking the difference on the hypothesis that C is constant, we have

$$\Delta u = C(-1)^x - \frac{x}{2} - \frac{1}{4} \ldots\ldots\ldots\ldots\ldots(b).$$

Taking again the difference of (a) on the hypothesis that C is variable, we have, on reduction,

$$\Delta u = C(-1)^x - \frac{x}{2} - \frac{1}{4}$$

$$+ 2C\Delta C(-1)^{2x} + (\Delta C)^2(-1)^{2x} + \frac{\Delta C}{2}(-1)^x \ldots\ldots (c),$$

which agrees in form with (b) if we have

$$\Delta C \left\{ 2C(-1)^{2x} + \Delta C(-1)^{2x} + \frac{1}{2}(-1)^x \right\} = 0.$$

Here the component equation $\Delta C = 0$ leads back to the given primitive; the remaining equation is

$$2C(-1)^{2x} + \Delta C(-1)^{2x} + \frac{1}{2}(-1)^x = 0 ;$$

or dividing by $(-1)^{2x}$, and transposing,

$$\Delta C + 2C = -\frac{1}{2}(-1)^{-x},$$

the integral of which is

$$C = \left(c + \frac{x}{2}\right)(-1)^{-x}.$$

Substituting this in the given primitive, we get

$$u = \left(c + \frac{x}{2} - \frac{1}{4}\right)^2 - \frac{x^2}{4}$$

$$= \left(c - \frac{1}{4}\right)^2 + \left(c - \frac{1}{4}\right)x,$$

or, replacing $c - \frac{1}{4}$ by c, since the constant is arbitrary,

$$u = cx + c^2,$$

the indirect integral sought.

Ex. 5. The equation $u_{x+1} = (1 + u_x^{\frac{1}{3}})^3$ is satisfied by

$$u_x = (x + c)^3 \dots\dots\dots\dots\dots (a);$$

deduce thence a cycle of three complete primitives.

By (4), Art. 3, the equation for determining c is

$$(x + 1 + Dc)^3 - (x + 1 + c)^3 = 0,$$

which is resolvable into

$$\Delta c = 0 \dots\dots\dots\dots\dots(b),$$

$$Dc - \mu c = (\mu - 1)(x + 1) \dots\dots\dots (c),$$

$$Dc - \nu c = (\nu - 1)(x + 1) \dots\dots\dots(d),$$

μ and ν being roots of the equation

$$\mu^2 + \mu + 1 = 0,$$

i. e. imaginary cube roots of unity. The equation (b) leads back to the given primitive, while (c) and (d) give on integration,

$$c = -x - \frac{\mu}{\mu - 1} + c_1 \mu^x,$$

$$c = -x - \frac{\nu}{\nu - 1} + c_2 \nu^x;$$

whence, substituting in (a), we have

$$u = \left(c_1 \mu^x - \frac{\mu}{\mu - 1} \right)^3 \dots\dots\dots\dots (e),$$

$$u = \left(c_2 \nu^x - \frac{\nu}{\nu - 1} \right)^3 \dots\dots\dots\dots (f);$$

and these with (a) form the cycle in question.

It will be found that either of the equations (e), (f), assumed as complete primitive, leads to the other and to (a) as indirect integrals.

The principle of Continuity.

4. We have defined the Calculus of Finite Differences as the science which is occupied about the ratios of the simultaneous increments of quantities mutually dependent; the Differential Calculus, as the science which is occupied about the limits to which such ratios approach as the increments are indefinitely diminished. The terms of these definitions forbid us to regard the Differential Calculus as merely a particular case of the Calculus of Finite Differences. And a careful analysis of the meaning of the word *limit* will shew us that it is not true that every result of the Calculus of Finite Differences merges when the increments are indefinitely diminished into a result of the Differential Calculus.

It is a familiar but a partial illustration which presents a curve as the limit to which a polygon tends as its sides are indefinitely increased in number and diminished in length. Let us suppose the differences of the value of the abscissa x for the successive points of the polygon to be constant, the law connecting the ordinates of these points to be expressed by an equation of differences, and the corresponding law of the ordinates of the limiting curve to be expressed by a differential equation.

Now there is a more complete and there is a less complete sense in which a curve may be said to be the limit of a polygon.

In the more complete sense not only does every angular point in the perimeter of the polygon approach in the transition to the limit indefinitely near to the curve, but every side of the polygon tends also indefinitely to *coincidence* with the curve. In virtue of this latter condition the value of $\frac{\Delta y}{\Delta x}$ in the polygon tends as Δx is diminished to that of $\frac{dy}{dx}$ in the curve. It is evident that this condition will be realized if the angles of the polygon in its state of transition are all salient, and tend to π as their limit.

But suppose the angles to be alternately salient and re-entrant, and, while the sides of the polygon are indefinitely diminished, to continue to be such without tending to any limit in which that character of alternation would cease. Here it is evident that while every point in the circumference of the polygon approaches indefinitely to the curve, its linear elements do not tend to coincidence of direction with the curve. Here then the limit to which $\frac{\Delta y}{\Delta x}$ approaches in the polygon is not the same as the value of $\frac{dy}{dx}$ in the curve.

If then the solutions of an equation of differences of the first order be represented by geometrical loci, and if, as Δx approaches to 0, these loci tend, some after the first, some after the second, of the above modes to continuous curves; then such of those curves as have resulted from the former process and are limits of their generating polygons in respect of the ultimate *direction* of the linear elements as well as position of their extreme points, will alone represent the solutions of the differential equations into which the equation of differences will have merged. This is the geometrical expression of the principle of continuity.

The principle admits also of analytical expression. Assuming h as the indeterminate increment of x, let y_x, y_{x+h}, y_{x+2h} be the ordinates of three consecutive points of the polygon, let ϕ be the angle which the straight line joining the first and second of these points makes with the axis of x, ψ the corresponding angle for the second and third of the points, and let $\psi - \phi$, or θ, be called the angle of contingence of these sides.

Now,

$$\tan \phi = \frac{y_{x+h} - y_x}{h}, \quad \tan \psi = \frac{y_{x+2h} - y_{x+h}}{h},$$

$$\tan \theta = \frac{\dfrac{y_{x+2h} - y_{x+h}}{h} - \dfrac{y_{x+h} - y_x}{h}}{1 + \dfrac{y_{x+h} - y_x}{h} \cdot \dfrac{y_{x+2h} - y_{x+h}}{h}}$$

$$= \frac{\dfrac{y_{x+2h} - 2y_{x+h} + h}{h}}{1 + \dfrac{y_{x+h} - y_x}{h} \dfrac{y_{x+2h} - y_{x+h}}{h}}.$$

Now, since $h = \Delta x$, we have,

$$y_{x+h} - y_x = \Delta y_x,$$
$$y_{x+2h} - 2y_{x+h} + y_x = \Delta^2 y_x,$$
$$y_{x+2h} - y_{x+h} = \Delta y_x + \Delta^2 y_x.$$

Therefore replacing y_x by y,

$$\tan \theta = \frac{\dfrac{\Delta^2 y_x}{\Delta x}}{1 + \left(\dfrac{\Delta y}{\Delta x}\right)^2 + \dfrac{\Delta y}{\Delta x}\dfrac{\Delta^2 y}{\Delta x}} \quad \text{............ (1)}.$$

Now the principle of continuity demands that in order that the solution of an equation of differences of the first order may merge into a solution of the limiting differential equation, the value which it gives to the above expression for $\tan \theta$ should, as Δx approaches to 0, tend to become infinitesimal; since in any continuous curve or continuous portion of a curve tan θ is infinitesimal. Again, that the above expression for tan θ should become infinitesimal, it is clearly necessary and sufficient that $\frac{\Delta^2 y}{\Delta x}$ should become so.

5. The application of this principle is obvious. Supposing that we are in possession of any of the complete primitives of an equation of differences in which Δx is indeterminate, then if in one of those primitives, the value of Δx being indefinitely diminished, that of $\frac{\Delta^2 y}{\Delta x}$ tends, *independently of the value of the arbitrary constant c*, to become infinitesimal also, the complete primitive merges into a complete primitive of the limiting differential equation; but if $\frac{\Delta^2 y}{\Delta x}$ tend to become infinitesimal with Δx only for a *particular* value of c, then only the *particular* integral corresponding to that value merges into a solution of the differential equation.

We have seen that when an equation of differences of the first order has two complete primitives standing in the mutual relation of direct and indirect integrals, each of them represents in geometry a system of envelopes to the loci represented by the other. Now suppose that one of these primitives should, according to the above process, merge into a complete primitive of the limiting differential equation, while the other furnishes only a particular solution; then the latter, not being included in the complete primitive of the differential equation, will be a singular solution, and retaining in the limit its geometrical character, will be a singular solution of the envelope species. Hence, the remarkable conclusion that those singular solutions of differential equations which are of the envelope species, originate from particular primitives of equations of differences; their isolation being due to the circumstance that the *associated* particular primitives of the equation of differences, not possessing that character which is required by the principle of continuity, are unrepresented in the solutions of the differential equation.

In the following examples we shall confine ourselves to those indirect integrals which arise from the supposition that the equation $\Delta c = 0$ is wholly rejected in the process of derivation from the given complete primitive, Art. 2. The other indirect integrals will, as is obvious from the mode of their formation, generally be of the purely discontinuous species.

Ex. 6. The differential equation $y = x\dfrac{dy}{dx} + \left(\dfrac{dy}{dx}\right)^2$ has for its complete primitive

$$y = cx + c^2 \dots\dots\dots\dots (a),$$

and for its singular solution, which is of the envelope species,

$$y = \frac{-x^2}{4} \dots\dots\dots\dots(b).$$

It is required to trace these back to their origin in the solution of an equation of differences. 1st, Taking the difference of the complete primitive, Δx being indeterminate and c a mere constant, we have

$$\Delta y = c\,\Delta x.$$

Hence $c = \dfrac{\Delta y}{\Delta x}$, and substituting in the complete primitive, we have

$$y = x\frac{\Delta y}{\Delta x} + \left(\frac{\Delta y}{\Delta x}\right)^2 \dots\dots\dots (c).$$

This is the equation of differences sought.

Taking the difference of (a), Δx being still indeterminate but c a variable parameter, we have

$$\Delta y = c\Delta x + x\Delta c + \Delta x\Delta c + 2c\Delta c + (\Delta c)^2,$$

which, equated with the previous value of Δy, gives on dividing by Δc

$$\Delta c + 2c = -(x + \Delta x),$$

an equation of differences for determining c.

To solve the equation, it is desirable to reduce it to the ordinary form in which the increment of the independent variable is unity. Let then $\Delta x = h$, and let $x = ht$, then

$$\Delta x = h\Delta t; \ \therefore\ \Delta t = 1.$$

And we have
$$\Delta c + 2c = -h(t+1),$$
the integral of which is
$$c = a(-1)^t - h\left(\frac{t}{2} + \frac{1}{4}\right)$$
$$= a(-1)^{\frac{x}{h}} - \frac{x}{2} - \frac{h}{4},$$
whence, substituting in (a),
$$y = \left\{a(-1)^{\frac{x}{h}} - \frac{h}{4}\right\}^2 - \frac{x^2}{4}$$
$$= a^2 - \frac{ha(-1)^{\frac{x}{h}}}{2} + \frac{h^2}{16} - \frac{x^2}{4} \dots\dots\dots\dots (d).$$

It results then that (c) has for complete primitives (a) and (d), h being equal to Δx.

2ndly. To determine $\tan\theta$ for the primitive (a), we have
$$\Delta y = c\Delta x, \quad \Delta^2 y = 0,$$
whence, substituting in (1), we find $\tan\theta = 0$. Thus the complete primitive (a) merges without limitation into a complete primitive of the differential equation.

But employing the complete primitive (d), we have
$$\Delta y = ha(-1)^{\frac{x}{h}} - \frac{2xh + h^2}{4},$$
$$\Delta^2 y = -2ha(-1)^{\frac{x}{h}} - \frac{h^2}{2}.$$
Hence
$$\frac{\Delta^2 y}{\Delta x} = -2a(-1)^{\frac{x}{h}} - \frac{h}{2}.$$

Now this value does not tend to 0 as h tends to 0, unless $a = 0$. Making therefore $a = 0$, $h = 0$, in (d), we have as the limiting value of y
$$y = -\frac{x^2}{4},$$
and this agrees with (b).

Thus while the complete primitive of the differential equation comes without any limitation of the arbitrary constant from the first complete primitive of the equation of differences, the singular solution of the differential equation is only the limiting form of a particular primitive included under the second of the complete primitives (d) of the equation of differences. Geometrically, that complete primitive represents a system of waving or zigzag lines, each of which perpetually crosses and recrosses some one of the system of parabolas represented by the equation

$$y = a^2 - \frac{x^2}{4}$$

As h tends to 0, those lines deviate to less and less distances on either side from the curves; but only one of these tends to ultimate *coincidence* with its limiting parabola.

When the given complete primitive of the differential equation is homogeneous with respect to x, y, and any constant other than of integration, it suffices to form the equation of differences on the hypothesis that $\Delta x = 1$, and examine the form which its solution assumes when the above quantities tend to become infinite, still retaining a finite ratio to each other. The following somewhat difficult problem illustrates this, and at the same time affords a valuable exercise in the treatment of functions of large numbers.

Ex. 7. The differential equation

$$y = x \frac{dy}{dx} + \frac{m}{\dfrac{dy}{dx}}$$

has for its complete primitive

$$y = cx + \frac{m}{c} \quad \dotfill \quad (a),$$

and for its singular solution

$$y^2 = 4mx.$$

It is required to trace these to their origin.

The complete primitive (a) is homogeneous with respect to y, x, and m, which it is therefore permitted to make infinite together, while c remains finite. Regarding Δx as 1, we are led to the equation of differences

$$y = x\Delta y + \frac{m}{\Delta y}.$$

To complete the cycle of primitives of this equation, we have from (a) in the usual way

$$c\,(c+\Delta c) = \frac{m}{x+1}.$$

To integrate this, take the logarithm of each member, and, putting $\log c = v_x$, we have

$$v_{x+1} + v_x = \log \frac{m}{x+1},$$

whence

$$v_x = (-1)^{x-1}\left\{ \Sigma \frac{1}{(-1)^x} \log \frac{m}{x+1} + \log a \right\},$$

$\log a$ being an arbitrary constant. Hence

$$v_x = \log \frac{m}{x} - \log \frac{m}{x-1} + \log \frac{m}{x-2} \cdots \pm \log \frac{m}{1} + (-1)^{x-1}\log a,$$

the last term but one being positive if x be odd, negative if x be even.

Hence, according as x is odd or even,

$$v_x = \log \frac{2.4\ldots(x-1)\,am}{1.3\ldots x} \text{ or } \log \frac{1.3\ldots(x-1)}{(2.4\ldots x)\,a}.$$

Therefore under the same restrictive conditions

$$c = \frac{2.4\ldots(x-1)\,am}{1.3\ldots x} \text{ or } \frac{1.3\ldots(x-1)}{(2.4\ldots x)\,a} \ldots\ldots\ldots (b).$$

Whence, by (a), according as x is odd or even,

$$y = \frac{2.4\ldots(x-1)\,am}{1.3\ldots(x-2)} + \frac{1.3\ldots x}{2.4\ldots(x-1)\,a},$$

$$y = \frac{1.3\ldots(x-1)}{2.4\ldots(x-2)\,a} + \frac{(2.4\ldots x)\,am}{1.3\ldots(x-1)},$$

these together constituting an indirect integral of the given equation.

Hence, putting for the present y_x for y regarded as a function of x, we have

$$y_{2x} = \frac{1 \cdot 3 \ldots (2x-1)}{2 \cdot 4 \ldots (2x-2)\,a} + \frac{(2 \cdot 4 \ldots 2x)\,am}{1 \cdot 3 \ldots (2x-1)},$$

$$y_{2x+1} = \frac{(2 \cdot 4 \ldots 2x)\,am}{1 \cdot 3 \ldots (2x-1)} + \frac{1 \cdot 3 \ldots (2x+1)}{(2 \cdot 4 \ldots 2x)\,a},$$

$$y_{2x+2} = \frac{1 \cdot 3 \ldots (2x+1)}{(2 \cdot 4 \ldots 2x)\,a} + \frac{2 \cdot 4 \ldots (2x+2)\,am}{1 \cdot 3 \ldots (2x+1)}.$$

Now if we suppose x large and reduce the factorials by the method illustrated in Ex. 5, Chap. VI., we find

$$y_{2x} = \left(\frac{2x^{\frac{1}{2}}}{\pi^{\frac{1}{2}}} - \frac{1}{4\pi^{\frac{1}{2}}x^{\frac{1}{2}}} \right) \frac{1}{a} + \left(\pi^{\frac{1}{2}}x^{\frac{1}{2}} + \frac{\pi^{\frac{1}{2}}}{8x^{\frac{1}{2}}} \right) am \ldots \ldots \ldots (c),$$

$$y_{2x+1} = \left(\pi^{\frac{1}{2}}x^{\frac{1}{2}} + \frac{\pi^{\frac{1}{2}}}{8x^{\frac{1}{2}}} \right) am + \left(\frac{2x^{\frac{1}{2}}}{\pi^{\frac{1}{2}}} + \frac{3}{4\pi^{\frac{1}{2}}x^{\frac{1}{2}}} \right) \frac{1}{a},$$

$$y_{2x+2} = \left(\frac{2x^{\frac{1}{2}}}{\pi^{\frac{1}{2}}} + \frac{3}{4\pi^{\frac{1}{2}}x^{\frac{1}{2}}} \right) \frac{1}{a} + \left(\pi^{\frac{1}{2}}x^{\frac{1}{2}} + \frac{5\pi^{\frac{1}{2}}}{8x^{\frac{1}{2}}} \right) ma.$$

Hence

$$y_{2x+1} - y_{2x} = \frac{1}{\pi^{\frac{1}{2}}x^{\frac{1}{2}}a},$$

$$y_{2x+2} - y_{2x+1} = \frac{\pi^{\frac{1}{2}}ma}{2x^{\frac{1}{2}}}.$$

Therefore $\Delta^2 y_{2x} = y_{2x+2} - 2y_{2x+1} + y_{2x}$

$$= \frac{\pi^{\frac{1}{2}}ma}{2x^{\frac{1}{2}}} - \frac{1}{\pi^{\frac{1}{2}}x^{\frac{1}{2}}a} \ldots \ldots \ldots \ldots \ldots (d).$$

Before we can deduce any conclusions from this expression we must determine of what order of magnitude a is to be considered.

B. F. D.   10

The general expression for y_{2x} in the new primitive is deduced from that of y in the original complete primitive by changing x into $2x$, and making

$$c = \frac{1.3 \ldots (2x-1)}{(2.4 \ldots 2x)\, a}.$$

Therefore

$$a = \frac{1.3 \ldots (2x-1)}{(2.4 \ldots 2x)\, c}$$

$$= \frac{1}{c \sqrt{(2\pi x)}},$$

x tending to infinity. Hence, since c is finite, a is of the same order of magnitude as $\dfrac{1}{\sqrt{x}}$. But m is of the same order as x. Hence the terms of the right-hand member of (d) are of such an order as to be finite, and the condition that that member must itself be infinitesimal gives

$$\frac{\pi^{\frac{1}{2}} m a}{2} - \frac{1}{\pi^{\frac{1}{2}} a} = 0,$$

whence

$$a = \sqrt{\left(\frac{2}{m\pi}\right)}.$$

Substituting this in (c) and reducing, we have

$$y_{2x} = 2\sqrt{(2mx)},$$

or

$$y_x = 2\sqrt{(mx)};$$

whence, restoring y for y_x,

$$y^2 = 4mx.$$

A similar result might be deduced by employing y_{2x+1} for y_{2x}.

Singular Solutions.

6. The following proposition, due to Poisson (*Journal de l'Ecole Polytechnique*, Tom. VI. p. 60), contains a theory of the singular solutions of equations of differences similar

to that of Laplace for differential equations. The student must particularly notice the *hypothesis* upon which the demonstration rests.

PROP. If an equation of differences of the first order of the form

$$\Delta y = f(x, y) \quad\ldots\ldots\ldots\ldots\ldots\ldots (1)$$

be satisfied by $y = u$, and if on developing $f(x, u + z)$ in ascending powers of z the index of z in the second term be less than unity, then $y = u$ will be a singular solution. But if that index be equal to or greater than unity, $y = u$ will be a particular integral.

Poisson begins by laying down the hypothesis that if $y = u$ be a particular integral obtained by making the arbitrary constant $C = a$ in a complete primitive, that primitive will be expressible in the form

$$y = u + (C - a)^\alpha X_1 + (C - a)^\beta X_2 \ldots,$$

$\alpha, \beta, \gamma \ldots$ being ascending, positive, and constant indices. He then proceeds to investigate a method by which the values of these indices and the forms of the functions X_1, X_2, &c. can be determined, regarding the failure of such method as an indication that the supposed particular primitive is a singular solution.

If we replace $(C - a)^\alpha$ by c the above equation assumes the form

$$y = u + cX_1 + c^\alpha X_2 + c^\beta X_3 + \&c. \ldots\ldots\ldots\ldots(2),$$

$\alpha, \beta,$ &c., being ascending, positive, and constant indices greater than unity.

Hence, substituting in (1)

$$\Delta u + c\Delta X_1 + c^\alpha \Delta X_2 + \&c. = f(x, u + cX_1 + c^\alpha X_2 \ldots),$$

or assuming

$$cX_1 + c^\alpha X_2 + \&c. = z,$$

$$\Delta u + c\Delta X_1 + c^\alpha \Delta X_2 \ldots = f(x, u + z) \ldots\ldots\ldots\ldots(3).$$

Now suppose the second member developable in the form

$$f(x, u) + f_1(x, u) z^m + f_2(x, u) z^n + \&c.,$$

the indices m, n, &c. being positive constants, the values of which will be known from the constitution of the function $f(x, u)$. Then replacing z by the series for which it stands, and observing that $\Delta u = f(x, u)$, (3), becomes

$$c\Delta X_1 + c^a \Delta X_2 + \&c. = f_1(x, u)(cX_1 + c^a X_2 \ldots)^m$$
$$+ f_2(x, u)(cX_1 + c^a X_2 \ldots)^n$$
$$+ \&c.,$$

or, developing the second member in ascending powers of c,

$$c\Delta X_1 + c^a \Delta X_2 \ldots = f_1(x, u) X_1{}^m c^m + \phi(x) c^{m'} + \&c. \ldots (4).$$

Now first let $m = 1$, then equating the first terms of the two members, we have

$$\Delta X_1 = f_1(x, u) X_1{}^m,$$

an equation of differences for determining X_1. This being found, we should have on equating the second terms of the members of (4) another equation of differences for determining X_2, and so on.

Secondly, let m be greater than 1; then making $\Delta X_1 = 0$, which determines X_1 as a constant, and assuming $a = m$, we get

$$\Delta X_2 = f_1(x, u) X_1{}^m,$$

which determines X_2, and so on. Here then also the particular integral can be completed.

But, thirdly, if m be less than 1 the two members of (4) cannot be made to agree. The supposed primitive cannot be completed, and is shewn to be a singular solution.

Poisson illustrates the above theory in the equation

$$\frac{y}{4} = \frac{4^{3x}(\Delta y)^3}{9} - \frac{\Delta y}{3} \ldots\ldots\ldots\ldots\ldots (a),$$

of which a complete primitive is

$$y = a\left(\frac{1}{2}\right)^{-2x} - \frac{3}{16} a^3,$$

and for which he obtains the singular solution

$$y = \pm \frac{8}{9} \left(-\frac{1}{2} \right)^{3x}.$$

The direct process involving the employment of Cardan's rule for determining Δy as a function of x and y being complicated, it is better to proceed as follows. Poisson's rule at once leads to the condition

$$\frac{d}{dy}(\Delta y) = \infty, \quad \text{or} \quad \frac{dy}{d\Delta y} = 0.$$

Applying this to (a) we get

$$(4)^{3x}(\Delta y)^2 - 1 = 0,$$

whence eliminating Δy by means of (a) we have the result in question. It must be ascertained by trial that it satisfies the equation (a).

If we attempted to deduce the above species of solutions directly from the complete primitive we should have to investigate the singular solution of the equation for determining c; so that the above process would still have to be employed.

Poisson says nothing about the geometrical character of these solutions. But it is clear from what has been shewn in the foregoing articles that they are not in any peculiar sense the primal forms of those singular solutions of differential equations which are geometrically interpreted as envelopes. They are of extremely rare occurrence, and I should not have deemed it worth while to notice them but for the double object of directing attention to a subject which seems to need further investigation, and of shewing how in those cases in which Poisson's hypothesis is realized a particular solution of an equation of differences may be completed.

EXERCISES.

1. Find a complete primitive of the equation

$$(\Delta y - a)(\Delta y - b) = 0,$$

which shall satisfy it by making $\Delta y = a$ for even, and $\Delta y = b$ for odd, values of x.

2. The equation

$$y = \frac{\Delta y}{2x + 1}\left(x^2 + \frac{\Delta y}{2x + 1}\right)$$

is satisfied by the complete primitive $y = cx^2 + c^2$. Shew that another complete primitive

$$y = \left\{a\,(-1)^x - \frac{x}{2}\right\}^2 - \frac{x^4}{4}$$

may thence be deduced.

3. Shew that a linear equation of differences admits of only one complete primitive.

4. The equation

$$\left(\frac{\Delta y}{a-1}\right)^2 + a^{2x}\frac{\Delta y}{a-1} - a^{2x}y = 0$$

has $y = ca^x + c^2$ for a complete primitive. Deduce another complete primitive.

5. In what sense are the complete primitives of an equation of differences of the first order said to be connected by a law of reciprocity?

CHAPTER IX.

1. THE symbolical methods for the solution of differential equations whether in finite terms or in series (*Diff. Equations,* Chap. XVII.) are equally applicable to the solution of equations of differences. Both classes of equations admit of the same symbolical form, the elementary symbols combining according to the same ultimate laws. And thus the only remaining difference is one of interpretation, and of processes founded upon interpretation. It is that kind of difference which exists between the symbols $\left(\dfrac{d}{dx}\right)^{-1}$ and Σ.

It has been shewn that if in a linear differential equation we assume $x = \epsilon^\theta$, the equation may be reduced to the form

$$f_0\left(\frac{d}{d\theta}\right) u + f_1\left(\frac{d}{d\theta}\right) \epsilon^\theta u + f_2\left(\frac{d}{d\theta}\right) \epsilon^{2\theta} u \ldots + f_n\left(\frac{d}{d\theta}\right) \epsilon^{n\theta} u = U,$$

$$\ldots\ldots\ldots (1),$$

U being a function of θ. Moreover, the symbols $\dfrac{d}{d\theta}$ and ϵ^θ obey the laws,

$$\left.\begin{aligned} f\left(\frac{d}{d\theta}\right) \epsilon^{m\theta} u &= \epsilon^{m\theta} f\left(\frac{d}{d\theta}+m\right) u \\ f\left(\frac{d}{d\theta}\right) \epsilon^{m\theta} &= f(m)\, \epsilon^{m\theta} \end{aligned}\right\} \ldots\ldots\ldots\ldots (2).$$

And hence it has been shewn to be possible, 1st, to express the solution of (1) in series, 2ndly, to effect by general theorems the most important transformations upon which finite integration depends.

Now $\dfrac{d}{d\theta}$ and ϵ^θ are the equivalents of $x\dfrac{d}{dx}$ and x, and it is proposed *to develope in this chapter the corresponding theory of equations of differences founded upon the analogous employment of the symbols* $x\dfrac{\Delta}{\Delta x}$ *and* xD *supposing* Δx *arbitrary, and therefore*

$$\Delta\phi\,(x) = \phi\,(x + \Delta x) - \phi\,(x),$$
$$D\phi\,(x) = \phi\,(x + \Delta x).$$

PROP. 1.   *If the symbols* π *and* ρ *be defined by the equations*

$$\pi = x\,\dfrac{\Delta}{\Delta x}\,,\ \ \rho = xD \ \dotsc\dotsc\dotsc\dotsc\dotsc (3),$$

they will obey the laws

$$\left.\begin{array}{c} f\,(\pi)\,\rho^m u = \rho^m f\,(\pi + m)\,u \\ f\,(\pi)\,\rho^m = f\,(m)\,\rho^m \end{array}\right\} \ \dotsc\dotsc\dotsc\dotsc (4),$$

the subject of operation in the second theorem being unity.

1st.   Let $\Delta x = r$, and first let us consider the interpretation of $\rho^m u_x$.

Now $\qquad\qquad \rho u_x = xD u_x = x u_{x+r}\,;$

$$\therefore\ \rho^2 u_x = \rho x u_{x+r} = x\,(x+r)\,u_{x+2r},$$

whence generally

$$\rho^m u_x = x(x+r)\,\dotsc\,\{x+(m-1)\,r\}\,u_{x+mr},$$

an equation to which we may also give the form

$$\rho^m u_x = x\,(x+r)\,\dotsc\,\{x+(m-1)\,r\}\,D^m u_x \ \dotsc\dotsc (5).$$

If $u_x = 1$, then, since $u_{x+mr} = 1$, we have

$$\rho^m 1 = x\,(x+r)\,\dotsc\,\{x+(m-1)\,r\},$$

to which we shall give the form

$$\rho^m = x\,(x+1)\,\dotsc\,\{x+(m-1)\,r\},$$

the subject 1 being understood.

2ndly. Consider now the series of expressions

$$\pi\rho^m u_x, \quad \pi^2\rho^m u_x, \quad \ldots \pi^n\rho^m u_x.$$

Now

$$\pi\rho^m u_x = x\,\frac{\Delta}{\Delta x}\,x\,(x+r)\,\ldots\,\{x+(m-1)\,r\}\,u_{x+mr}$$

$$= x\,\frac{(x+r)\,\ldots\,(x+mr)\,u_{x+(m+1)r} - x\,\ldots\,\{x+(m-1)r\}\,u_{x+mr}}{r}$$

$$= x\,\ldots\,\{x+(m-1)\,r\}\,\frac{(x+mr)\,u_{x+(m+1)r} - xu_{x+mr}}{r}$$

$$= x\,\ldots\,\{x+(m-1)\,r\}\,D^m\,\frac{xu_{x+r} - (x-mr)\,u_x}{r}$$

$$= \rho^m\,\frac{xu_{x+r} - (x-mr)\,u_x}{r}, \text{ by (5),}$$

$$= \rho^m\left(x\,\frac{u_{x+r} - u_x}{r} + mu_x\right)$$

$$= \rho^m\left(x\,\frac{\Delta}{\Delta x}\,u_x + mu_x\right)$$

$$= \rho^m\,(\pi+m)\,u_x.$$

Hence

$$\pi^2\rho^m u_x = \pi\rho^m\,(\pi+m)\,u_x$$
$$= \rho^m\,(\pi+m)^2 u_x,$$

and generally

$$\pi^n\rho^m u_x = \rho^m\,(\pi+m)^n u_x.$$

Therefore supposing $f(\pi)$ a function expressible in ascending powers of π, we have

$$f(\pi)\,\rho^m u = \rho^m f(\pi+m)\,u \ldots\ldots\ldots\ldots\ldots(6),$$

which is the first of the theorems in question.

Again, supposing $u = 1$, we have

$$f(\pi)\,\rho^m 1 = \rho^m f(\pi+m)\,1$$

$$= \rho^m\left\{f(m) + f'(m)\,\pi + \frac{f''(m)}{1.2}\,\pi^2 + \&c.\right\}1.$$

But $\pi 1 = x\dfrac{\Delta}{\Delta x}1 = 0,\quad \pi^2 1 = 0,$ &c.   Therefore

$$f(\pi)\rho^m 1 = \rho^m f(m)\,1.$$

Or, omitting but leaving understood the subject unity,

$$f(\pi)\rho^m = f(m)\,\rho^m \quad \dots \dots \dots \dots (7).$$

PROP. 2. *Adopting the previous definitions of π and ρ, every linear equation of differences admits of symbolical expression in the form*

$$f_0(\pi)u_x + f_1(\pi)\rho u_x + f_2(\pi)\rho^2 u_x \dots + f_m(\pi)\rho^m u_x = X \dots (8).$$

The above proposition is true irrespectively of the particular value of Δx, but the only cases which it is of any importance to consider are those in which $\Delta x = 1$ and -1.

First suppose the given equation of differences to be

$$X_0 u_{x+n} + X_1 u_{x+n-1} \dots + X_n u_x = \phi(x) \quad \dots \dots \dots (9).$$

Here it is most convenient to assume $\Delta x = 1$ in the expressions of π and ρ. Now multiplying each side of (9) by

$$x(x+1)\dots(x+n-1),$$

and observing that by (5)

$$xu_{x+1} = \rho u_x,\quad x(x+1)u_{x+2} = \rho^2 u_x,\ \&c.,$$

we shall have a result of the form

$$\phi_0(x)u_x + \phi_1(x)\rho u_x \dots + \phi_n(x)\rho^n u_x = \phi_1(x) \dots (10).$$

But since $\Delta x = 1,$

$$\pi = x\Delta,\quad \rho = xD$$
$$= x\Delta + x.$$

Hence

$$x = -\pi + \rho,$$

and therefore

$$\phi_0(x) = \phi_0(-\pi + \rho),\quad \phi_1(x) = \phi_1(-\pi + \rho),\ \&c.$$

These must be expressed in ascending powers of ρ, regard being paid to the law expressed by the first equation of (4).

The general theorem for this purpose, though its application can seldom be needed, is

$$F_0(\pi - \rho) = F_0(\pi) - F_1(\pi)\rho + F_2(\pi)\frac{\rho^2}{1.2}$$

$$- F_3(\pi)\frac{\rho^3}{1.2.3} + \&c. \dots\dots\dots (11),$$

where $F_1(\pi)$, $F_2(\pi)$, &c., are formed by the law

$$F_m(\pi) = F_{m-1}(\pi) - F_{m-1}(\pi - 1).$$

(*Diff. Equations*, p. 439.)

The equation (10) then assumes after reduction the form (8).

Secondly, suppose the given equation of differences presented in the form

$$X_0 u_x + X_1 u_{x-1} \dots + X_n u_{x-n} = X \dots\dots\dots (12).$$

Here it is most convenient to assume $\Delta x = -1$ in the expression of π and ρ.

Now multiplying (12) by $x(x-1)\dots(x-n+1)$, and observing that by (5)

$$x u_{x-1} = \rho u_x, \quad x(x-1)u_{x-2} = \rho^2 u_x, \&c.,$$

the equation becomes

$$\phi_0(x)u_x + \phi_1(x)\rho u_x \dots + \phi_n(x)\rho^n u_x = X,$$

but in this case as is easily seen we have

$$x = \pi + \rho,$$

whence, developing the coefficients, if necessary, by the theorem

$$F_0(\pi + \rho) = F_0(\pi) + F_1(\pi)\rho + F_2(\pi)\frac{\rho^2}{1.2} + \&c. \dots (13),$$

where as before

$$F_m(\pi) = F_{m-1}(\pi) - F_{m-1}(\pi - 1),$$

we have again on reduction an equation of the form (8).

2. It is not always necessary in applying the above methods of reduction to multiply the given equation by a factor of the form

$$x (x + 1) \ldots (x + n - 1), \text{ or } x (x - 1) \ldots (x - n + 1),$$

to prepare it for the introduction of ρ. It may be that the constitution of the original coefficients $X_0, X_1 \ldots X_n$ is such as to render this multiplication unnecessary; or the requisite factors may be introduced in another way. Thus resuming the general equation

$$X_0 u_x + X_1 u_{x-1} \ldots + X_n u_{x-n} = 0 \ldots\ldots\ldots(14),$$

assume

$$u_x = \frac{v_x}{1 . 2 \ldots x}.$$

We find

$$X_0 v_x + X_1 x v_{x-1} \ldots + X_n x (x - 1) \ldots (x - n + 1) v_{x-n} = 0 \ldots (15).$$

Hence assuming

$$\pi = x \frac{\Delta}{\Delta x}, \quad \rho = xD,$$

where $\Delta x = -1$, we have

$$X_0 v_x + X_1 \rho v_x \ldots + X_n \rho^n v_x = 0 \ldots\ldots\ldots(16),$$

and it only remains to substitute $\pi + \rho$ for x and develope the coefficients by (13).

3. A preliminary transformation which is often useful consists in assuming $u_x = \mu^x v_x$. This converts the equation

$$X_0 u_x + X_1 u_{x-1} \ldots + X_n u_{x-n} = 0 \ldots\ldots\ldots(17)$$

into

$$\mu^n X_0 v_x + \mu^{n-1} X_1 v_{x-1} \ldots X_n v_{x-n} = 0 \ldots\ldots(18),$$

putting us in possession of a disposable constant μ.

4. When the given equation of differences is expressed directly in the form

$$X_0 \Delta^n u + X_1 \Delta^{n-1} u \ldots + X_n u = 0 \ldots\ldots\ldots(19),$$

it may be convenient to apply the following theorem.

Theorem. If $\pi = x \dfrac{\Delta}{\Delta x}$, $\rho = xD$, then

$$\pi (\pi - 1) \ldots (\pi - n + 1) u = x (x + \Delta x) \ldots$$

$$\{x + (n - 1) \Delta x\} \left(\frac{\Delta}{\Delta x}\right)^n u \ldots\ldots\ldots\ldots (20).$$

To prove this we observe that since

$$F(\pi) \rho^n u = \rho^n F(\pi + n) u,$$

therefore $F(\pi + n) u = \rho^{-n} F(\pi) \rho^n u,$

whence $F(\pi - n) u = \rho^n F(\pi) \rho^{-n} u.$

Now reversing the order of the factors $\pi, \pi - 1, \ldots \pi - n + 1$ in the first member of (20) and applying the above theorem to each factor separately, we have

$$(\pi - n + 1) (\pi - n + 2) \ldots \pi u$$

$$= \rho^{n-1} \pi \rho^{-n+1} \rho^{n-2} \pi \rho^{-n+2} \ldots \pi u$$

$$= \rho^n (\rho^{-1} \pi)^n u.$$

But $\rho^{-1} \pi = (xD)^{-1} x \dfrac{\Delta}{\Delta x} = D^{-1} x^{-1} x \dfrac{\Delta}{\Delta x} = D^{-1} \dfrac{\Delta}{\Delta x};$

$$\therefore (\pi - n + 1) (\pi - n + 2) \ldots \pi = \rho^n \left(D^{-1} \frac{\Delta}{\Delta x}\right)^n$$

$$= \rho^n D^{-n} \left(\frac{\Delta}{\Delta x}\right)^n.$$

But $\rho^n u = x (x + r) \ldots \{x + (n - 1) r\} D^n u$, whence

$$(\pi - n + 1)(\pi - n + 2) \ldots \pi u = x (x + r) \ldots \{x + (n - 1) r\} \left(\frac{\Delta}{\Delta x}\right)^n u,$$

which, since $r = \Delta x$, agrees with (20).

When $\Delta x = 1$, the above gives

$$\pi (\pi - 1) \ldots (\pi - n + 1) = x (x + 1) \ldots (x + n - 1) \Delta^n \ldots (21).$$

Hence, resuming (19), multiplying both sides by

$$x (x + 1) \ldots (x + n - 1),$$

and transforming, we have a result of the form

$$\phi_0(x)\,\pi\,(\pi-1)\,...\,(\pi-n+1)\,u$$
$$+\phi_1(x)\,\pi\,(\pi-1)\,...\,(\pi-n+2)\,u+\&\text{c.}=0.$$

It only remains then to substitute $x=-\pi+\rho$, develope the coefficients, and effect the proper reductions.

Solution of Linear Equations of Differences in series.

5. Supposing the second member 0, let the given equation be reduced to the form

$$f_0(\pi)\,u+f_1(\pi)\,\rho u+f_2(\pi)\,\rho^2 u\,...\,+f_n(\pi)\,\rho^n u=0.....(22),$$

and assume $u=\Sigma a_m\rho^m$. Then substituting and attending to the first equation of (4), we have

$$\Sigma\{f_0(\pi)\,a_m\rho^m+f_1(\pi)\,a_m\rho^{m+1}\,...\,+f_n(\pi)\,a_m\rho^{m+n}\}=0,$$

whence, by the second equation of (4),

$$\Sigma\{f_0(m)\,a_m\rho^m+f_1(m+1)\,a_m\rho^{m+1}\,...\,+f_n(m+n)\,a_m\rho^{m+n}\}=0,$$

in which the aggregate coefficient of ρ^m equated to 0 gives

$$f_0(m)\,a_m+f_1(m)\,a_{m-1}\,...\,+f_n(m)\,a_{m-n}=0.........(23).$$

This, then, is the relation connecting the successive values of a_m. The lowest value of m, corresponding to which a_m is arbitrary, will be determined by the equation

$$f_0(m)=0,$$

and there will thus be as many values of u expressed in series as the equation has roots.

If in the expression of π and ρ we assume $\Delta x=1$, then since

$$\rho^m=x\,(x+1)\,...\,(x+m-1)...............(24),$$

the series $\Sigma a_m \rho^m$ will be expressed in ascending factorials of the above form. But if in expressing π and ρ we assume $\Delta x = -1$, then since

$$\rho^m = x(x-1) \dots (x-m+1)\dots\dots\dots\dots(25),$$

the series will be expressed in factorials of the latter form.

Ex. 1. Given

$$(x-a) u_x - (2x-a-1) u_{x-1} + (1-q^2)(x-1) u_{x-2} = 0 ;$$

required the value of u_x in descending factorials.

Multiplying by x, and assuming $\pi = x \dfrac{\Delta}{\Delta x}$, $\rho = xD$, where $\Delta x = -1$, we have

$$x(x-a) u_x - (2x-a-1) \rho u_x + (1-q^2) \rho^2 u_x = 0,$$

whence, substituting $\pi + \rho$ for x, developing by (11), and reducing,

$$\pi(\pi-a) u_x - q^2 \rho^2 u_x = 0\dots\dots\dots\dots\dots(a).$$

Hence $u_x = \Sigma a_m \rho^m,$

the initial values of a_m corresponding to $m = 0$ and $m = a$ being arbitrary, and the succeeding ones determined by the law

$$m(m-a) a_m - q^2 a_{m-2} = 0.$$

Thus we have for the complete solution

$$u_x = C \left\{ 1 + \frac{q^2 x^{(2)}}{2(2-a)} + \frac{q^4 x^{(4)}}{2.4.(2-a).(4-a)} + \&c. \right\}$$

$$+ C' \left\{ x^{(a)} + \frac{q^2 x^{(a+2)}}{2.(2+a)} + \frac{q^4 x^{(a+4)}}{2.4(2+a)(4+a)} + \&c. \right\}\dots(b).$$

It may be observed that the above equation of differences might be so prepared that the complete solution should admit of expression in finite series. For assuming $u_x = \mu^x v_x$, and then transforming as before, we find

$$\mu^2\pi\,(\pi-a)\,v_x + (\mu^2-\mu)\,(2\pi-a-1)\,\rho v_x$$
$$+ \{(\mu-1)^2 - q^2\}\,\rho^2 v_x = 0 \dots\dots\dots (c),$$

which becomes binomial if $\mu = 1 \pm q$, thus giving

$$\pi\,(\pi-a)\,v_x + \frac{\mu-1}{\mu}\,(2\pi-a-1)\,\rho v_x = 0.$$

Hence we have for either value of μ,

$$u_x = \mu^x \Sigma a_m \rho^m = \mu^x \Sigma a_m x\,(x-1)\,\dots\,(x-m+1)\dots\dots(d),$$

the initial value of m being 0 or a, and all succeeding values determined by the law

$$m\,(m-a)\,a_m + \frac{\mu-1}{\mu}\,(2m-a-1)\,a_{m-1} = 0 \dots\dots(e).$$

It follows from this that the series in which the initial value of m is 0 terminates when a is a positive odd number, and the series in which the initial value of m is a terminates when a is a negative odd number. Inasmuch however as there are two values of μ, either series, by giving to μ both values in succession, puts us in possession of the complete integral.

Thus in the particular case in which a is a positive odd number we find

$$u_x = C\,(1+q)^x \left\{ 1 - \frac{q}{1+q}\frac{(1-a)\,x}{1\,.\,(1-a)} \right.$$
$$\left. + \frac{q^2}{(1+q)^2}\frac{(1-a)\,(3-a)\,x^{(2)}}{1\,.\,2\,(1-a)\,(2-a)} - \&c. \right\}$$
$$+ C'\,(1-q)^x \left\{ 1 + \frac{q}{1+q}\frac{(1-a)\,x}{1\,.\,(1-a)} \right.$$
$$\left. + \frac{q^2}{(1+q)^2}\frac{(1-a)\,(3-a)\,x^{(2)}}{1\,.\,2\,(1-a)\,(3-a)} + \&c. \right\} \dots(h).$$

The above results may be compared with those of p. 454 of *Differential Equations*.

Finite solution of Equations of Differences.

6. The simplest case which presents itself is when the symbolical equation (8) is monomial, i. e. of the form

$$f_0(\pi) u = X \dots\dots\dots\dots\dots (26).$$

We have thus

$$u = \{f_0(\pi)\}^{-1} X \dots\dots\dots\dots (27).$$

Resolving then $\{f_0(\pi)\}^{-1}$ as if it were a rational algebraic fraction, the complete value of u will be presented in a series of terms of the form

$$A(\pi - a)^{-i} X.$$

But by (4) we have

$$(\pi - a)^{-i} X = \rho^a (\pi)^{-i} \rho^{-a} X \dots\dots\dots (28).$$

It will suffice to examine in detail the case in which $\Delta x = 1$ in the expression of π and ρ.

To interpret the second member of (28) we have then

$$\rho^a \phi(x) = x(x+1) \dots (x+a-1) \phi(x+a),$$

$$\rho^{-a} \phi(x) = \frac{\phi(x-a)}{(x+1)(x+2) \dots (x+a)},$$

$$\pi^{-i} \phi(x) = (x\Delta)^{-i} \phi(x)$$

$$= \Sigma \frac{1}{x} \Sigma \frac{1}{x} \dots \phi(x);$$

the complex operation $\Sigma \dfrac{1}{x}$, denoting division of the subject by x and subsequent integration, being repeated i times.

Should X however be rational and integral it suffices to express it in factorials of the forms

$$x, \quad x(x+1), \quad x(x+1)(x+2), \quad \&c.$$

to replace these by ρ, ρ^2, ρ^3, &c. and then interpret (27) at once by the theorem

$$\{f_0(\pi)\}^{-1}\rho^m = \{f_0(m)\}^{-1}\rho^m$$
$$= \{f_0(m)\}^{-1}x(x+1)\ldots(x+m-1)\ldots(29).$$

As to the complementary function it is apparent from (28) that we have

$$(\pi - a)^{-i}0 = \rho^a\pi^{-i}0.$$

Hence in particular if $i = 1$, we find

$$(\pi - a)^{-1}0 = \rho^a\pi^{-1}0$$
$$= \rho^a\Sigma x^{-1}0$$
$$= C\rho^a$$
$$= Cx(x+1)\ldots(x+a-1)\ldots\ldots(30).$$

This method enables us to solve any equation of the form

$$x(x+1)\ldots(x+n-1)\Delta^n u + A_1 x(x+1)\ldots$$
$$\ldots(x+n-2)\Delta^{n-1}u\ldots+A_n u = X\ldots\ldots(31).$$

For symbolically expressed any such equation leads to the monomial form

$$\{\pi(\pi-1)\ldots(\pi-n+1)+A_1\pi(\pi-1)\ldots$$
$$\ldots(\pi-n+2)\ldots+A_n\}u = X\ldots\ldots\ldots(32).$$

Ex. 2.   Given

$$x(x+1)\Delta^2 u - 2x\Delta u + 2u = x(x+1)(x+2).$$

The symbolical form of this equation is

$$\pi(\pi-1)u - 2\pi u + 2u = x(x+1)(x+2)\ldots\ldots(a),$$

or $$(\pi^2 - 3\pi + 2)u = \rho^3.$$

Hence $$u = (\pi^2 - 3\pi + 2)^{-1}\rho^3$$
$$= (3^2 - 3\times 3 + 2)^{-1}\rho^3$$
$$+ C_1\rho^2 + C_2\rho,$$

since the factors of $\pi^2 - 3\pi + 2$ are $\pi - 2$ and $\pi - 1$. Thus we have

$$u = \frac{x\,(x+1)\,(x+2)}{2} + C_1 x\,(x+1) + C_2 x \ldots \ldots (b).$$

Binomial Equations.

7. Let us next suppose the given equation binomial and therefore susceptible of reduction to the form

$$u + \phi\,(\pi)\,\rho^n u = U \ldots \ldots \ldots \ldots \ldots (33),$$

in which U is a known, u the unknown and sought function of x. The possibility of finite solution will depend upon the form of the function $\phi\,(\pi)$, and its theory will consist of two parts, the first relating to the conditions under which the equation is directly resolvable into equations of the first order, the second to the laws of the transformations by which equations not obeying those conditions may when possible be reduced to equations obeying those conditions.

As to the first point it may be observed that if the equation be

$$u + \frac{1}{a\pi + b}\,\rho u = U \ldots \ldots \ldots \ldots (34),$$

it will, on reduction to the ordinary form, be integrable as an equation of the first order.

Again, if in (33) we have

$$\phi\,(\pi) = \psi\,(\pi)\,\psi\,(\pi - 1)\,\ldots\,\psi\,(\pi - n + 1),$$

in which $\psi\,(\pi) = \frac{1}{a\pi + b}$, the equation will be resolvable into a system of equations of the first order. This depends upon the general theorem that the equation

$$u + a_1\phi\,(\pi)\,\rho u + a_2\phi\,(\pi)\,\phi\,(\pi - 1)\,\rho^2 u \ldots$$
$$+ a_n\phi\,(\pi)\,\phi\,(\pi - 1)\,\ldots\,\phi\,(\pi - n + 1)\,\rho^n u = U$$

may be resolved into a system of equations, of the form

$$u - q\phi(\pi)\rho u = U,$$

q being a root of the equation

$$q^n + a_1 q^{n-1} + a_2 q^{n-2} \ldots + a_n = 0.$$

(*Differential Equations*, p. 405.)

Upon the same principle of formal analogy the propositions upon which the transformation of differential equations depends (*Ib.* pp. 408-9) might be adopted here with the mere substitution of π and ρ for D and ϵ^θ. But we prefer to investigate what may perhaps be considered as the most general forms of the theorems upon which these propositions rest.

From the binomial equation (33), expressed in the form

$$\{1 + \phi(\pi)\rho^n\}u = U,$$

we have

$$u = \{1 + \phi(\pi)\rho^n\}^{-1} U,$$

and this is a particular case of the more general form,

$$u = F\{\phi(\pi)\rho^n\} U \ldots\ldots\ldots\ldots\ldots (35).$$

Thus the unknown function u is to be determined from the known function U by the performance of a particular operation of which the *general* type is

$$F\{\phi(\pi)\rho^n\}.$$

Now suppose the given equations transformed by some process into a new but integrable binomial form,

$$v + \psi(\pi)\rho^n v = V,$$

V being here the given and v the sought function of x. We have

$$v = \{1 + \psi(\pi)\rho^n\}^{-1} V,$$

which is a particular case of $F\{\psi(\pi)\rho^n\} V$, supposing $F(t)$ to denote a function developable by Maclaurin's theorem. It is apparent therefore that the theory of this transformation must depend upon the theory of the connexion of the forms,

$$F\{\phi(\pi)\,\rho^n\}, \quad F\{\psi(\pi)\,\rho^n\}.$$

Let then the following inquiry be proposed. Given the forms of $\phi(\pi)$ and $\psi(\pi)$, is it possible to determine an operation $\chi(\pi)$ such that we shall have generally

$$F\{\phi(\pi)\,\rho^n\}\,\chi(\pi)\,X = \chi(\pi)\,F\{\psi(\pi)\,\rho^n\}\,X \dots\dots (36),$$

irrespectively of the form of X?

Supposing $F(t) = t$, we have to satisfy

$$\phi(\pi)\,\rho^n\chi(\pi)\,X = \chi(\pi)\,\psi(\pi)\,\rho^n\,X \dots\dots\dots (37).$$

Hence by the first equation of (4),

$$\phi(\pi)\,\chi(\pi - n)\,\rho^n X = \psi(\pi)\,\chi(\pi)\,\rho^n X,$$

to satisfy which, independently of the form of X, we must have

$$\psi(\pi)\,\chi(\pi) = \phi(\pi)\,\chi(\pi - n)\,;$$

$$\therefore \chi(\pi) = \frac{\phi(\pi)}{\psi(\pi)}\,\chi(\pi - n).$$

Therefore solving the above equation of differences,

$$\chi(\pi) = CP_n\,\frac{\phi(\pi)}{\psi(\pi)}.$$

Substituting in (37), there results,

$$\phi(\pi)\,\rho^n P_n\,\frac{\phi(\pi)}{\psi(\pi)}\,X = P_n\,\frac{\phi(\pi)}{\psi(\pi)}\,\psi(\pi)\,\rho^n X,$$

or, replacing $P_n\,\dfrac{\phi(\pi)}{\psi(\pi)}\,X$ by X_1,

and therefore X by $\left\{P_n\,\dfrac{\phi(\pi)}{\psi(\pi)}\right\}^{-1} X_1$,

$$\phi(\pi)\,\rho^n X_1 = P_n\,\frac{\phi(\pi)}{\psi(\pi)}\,\psi(\pi)\,\rho^n\left\{P_n\,\frac{\phi(\pi)}{\psi(\pi)}\right\}^{-1} X_1.$$

If for brevity we represent $P_n \dfrac{\phi(\pi)}{\psi(\pi)}$ by P, and drop the suffix from X_1 since the function is arbitrary, we have

$$\phi(\pi) \rho^n X = P\psi(\pi) \rho^n P^{-1} X.$$

Hence therefore

$$\{\phi(\pi) \rho^n\}^2 X = P\psi(\pi) \rho^n P^{-1} P\psi(\pi) \rho^n P^{-1} X$$
$$= P\{\psi(\pi) \rho^n\}^2 P^{-1}X,$$

and continuing the process,

$$\{\phi(\pi) \rho^n\}^m X = P\{\psi(\pi) \rho^n\}^m P^{-1} X.$$

Supposing therefore $F(t)$ to denote any function developable by Maclaurin's theorem, we have

$$F\{\phi(\pi) \rho^n\}X = PF\{\psi(\pi) \rho^n\} P^{-1}X.$$

We thus arrive at the following theorem.

THEOREM. *The symbols π and ρ combining in subjection to the law*

$$f(\pi) \rho^m X = \rho^m f(\pi + m) X,$$

the members of the following equation are symbolically equivalent, viz.

$$F\{\phi(\pi) \rho^n\} = P_n \frac{\phi(\pi)}{\psi(\pi)} F\{\psi(\pi) \rho^n\} P_n \frac{\psi(\pi)}{\phi(\pi)} \ \ldots \ (38).$$

A. From this theorem it follows, in particular, that we can always convert the equation

$$u + \phi(\pi) \rho^n u = U,$$

into any other binomial form,

$$v + \psi(\pi) \rho^n v = P_n \frac{\psi(\pi)}{\phi(\pi)} U,$$

by assuming $u = P_n \dfrac{\phi(\pi)}{\psi(\pi)} v.$

For we have

$$u = \{1 + \phi\,(\pi)\,\rho^n\}^{-1}\,U$$

$$= P_n\,\frac{\phi\,(\pi)}{\psi\,(\pi)}\,\{1 + \psi\,(\pi)\,\rho^n\}^{-1}\,P_n\,\frac{\psi\,(\pi)}{\phi\,(\pi)}\,U,$$

whence since

$$v = \{1 + \psi\,(\pi)\,\rho^n\}^{-1}\,V,$$

it follows that we must have

$$V = P_n\,\frac{\psi\,(\pi)}{\phi\,(\pi)}\,U, \quad u = P_n\,\frac{\phi\,(\pi)}{\psi\,(\pi)}\,v.$$

In applying the above theorem, it is of course necessary that the functions $\phi\,(\pi)$ and $\psi\,(\pi)$ be so related that the continued product denoted by $P_n\,\dfrac{\phi\,(\pi)}{\psi\,(\pi)}$ should be finite. The conditions relating to the introduction of arbitrary constants have been stated with sufficient fulness elsewhere, (*Differential Equations*, Chap. XVII. Art. 4).

B. The reader will easily demonstrate also the following theorem, viz:

$$F\{\phi\,(\pi)\,\rho^n\}\,X = \rho^m\,F\{\phi\,(\pi + m)\,\rho^n\}\,\rho^{-m}\,X,$$

and deduce hence the consequence that the equation

$$u + \phi\,(\pi)\,\rho^n u = U,$$

may be converted into

$$v + \phi\,(\pi + m)\,\rho^n v = \rho^{-m}\,U,$$

by assuming $u = \rho^m v$.

8. These theorems are in the following sections applied to the solution, or rather to the discovery of the conditions of finite solution, of certain classes of equations of considerable generality. In the first example the second member of the given equation is supposed to be any function of x. In the two others it is supposed to be 0. But the conditions of finite solution, if by this be meant the reduction of the discovery of the unknown quantity to the performance of a finite

number of operations of the kind denoted by Σ, will be the same in the one case as in the other. It is however to be observed, that when the second member is 0, a finite integral may be frequently obtained by the process for solutions in series developed in Art. 5, while if the second member be X, it is almost always necessary to have recourse to the transformations of Art. 7.

Discussion of the equation

$$(ax + b)\, u_x + (cx + e)\, u_{x-1} + (fx + g)\, u_{x-2} = X \ldots\ldots (a).$$

Consider first the equation

$$(ax + b)\, u_x + (cx + e)\, u_{x-1} + f\,(x - 1)\, u_{x-2} = X \ldots\ldots (b).$$

Let $u_x = \mu^x v_x$, then, substituting, we have

$$\mu^2 \,(ax + b)\, v_x + \mu\, (cx + e)\, v_{x-1} + f\, (x - 1)\, v_{x-2} = \mu^{-x+2} X.$$

Multiply by x and assume $\pi = x \dfrac{\Delta}{\Delta x}$, $\rho = xD$, in which $\Delta x = -1$, then

$$\mu^2 \,(ax^2 + bx)\, v_x + \mu\, (cx + e)\, \rho v_x + f\rho^2 v_x = x\mu^{-x+2} X,$$

whence, substituting $\pi + \rho$ for x and developing the coefficients, we find

$$\mu^2 \,(a\pi^2 + b\pi)\, v_x + \mu\, \{(2a\mu + c)\, \pi + (b - a)\, \mu + e\}\, \rho v_x$$
$$+ \,(a\mu^2 + c\mu + f)\, \rho^2 v_x = x\mu^{-x+2} X \ldots\ldots\ldots\ldots (c),$$

and we shall now seek to determine μ so as to reduce this equation to a binomial form.

1st. Let μ be determined by the condition

$$a\mu^2 + c\mu + f = 0,$$

then making

$$2a\mu + c = A, \quad (b - a)\,\mu + e = B,$$

we have

$$a\mu\pi \left(\pi + \frac{b}{a}\right) v_x + A\left(\pi + \frac{B}{A}\right) \rho v_x = x\mu^{-x+1} X,$$

or

$$v_x + \frac{A}{a\mu} \frac{\pi + \dfrac{B}{A}}{\pi \left(\pi + \dfrac{b}{a}\right)} \rho v_x = \frac{1}{a\mu} \left\{ \pi \left(\pi + \frac{b}{a}\right)\right\}^{-1} x\mu^{-x+1} X,$$

or, supposing V to be any *particular* value of the second member obtained by Art. 6, for it is not necessary at this stage to introduce an arbitrary constant,

$$v_x + \frac{A}{a\mu} \frac{\pi + \dfrac{B}{A}}{\pi \left(\pi + \dfrac{b}{a}\right)} \rho v_x = V \ldots\ldots\ldots\ldots (d).$$

This equation can be integrated when either of the functions,

$$\frac{B}{A}, \quad \frac{B}{A} - \frac{b}{a},$$

is an integer. In the former case we should assume

$$w_x + \frac{A}{a\mu} \frac{1}{\pi + \dfrac{b}{a}} \rho w_x = W_x \ldots\ldots\ldots\ldots\ldots (e),$$

whence we should have by (A),

$$v_x = P_1 \frac{\pi + \dfrac{B}{A}}{\pi} w_x, \quad W_x = P_1 \frac{\pi}{\pi + \dfrac{B}{A}} V \ldots\ldots\ldots\ldots (f).$$

In the latter case we should assume as the transformed equation

$$w_x + \frac{A}{a\mu} \frac{1}{\pi} \rho w_x = W_x \ldots\ldots\ldots\ldots\ldots (g),$$

and should find

$$v_x = P_1 \frac{\pi + \dfrac{B}{A}}{\pi + \dfrac{b}{a}} w_x, \quad W_x = P_1 \frac{\pi + \dfrac{b}{a}}{\pi + \dfrac{B}{A}} V \ldots\ldots\ldots (h).$$

The value of W_x obtained from (f) or (h) is to be substituted in (e) or (g), w_x then found by integration, and v_x determined by (f) or (h). One arbitrary constant will be introduced in the integration for w_x, and the other will be due either to the previous process for determining W_x, or to the subsequent one for determining v_x.

Thus in the particular case in which $\dfrac{B}{A}$ is a positive integer, we should have

$$W_x = \left\{\left(\pi + \frac{B}{A}\right)\left(\pi + \frac{B}{A} - 1\right) \ldots (\pi + 1)\right\}^{-1} 0,$$

a particular value of which, derived from the interpretation of $\left(\pi + \dfrac{B}{A}\right)^{-1} 0$ and involving an arbitrary constant, will be found to be $\dfrac{c}{1 + x}$. Substituting in (e) and reducing the equation to the ordinary unsymbolical form, we have,

$$\mu (ax + b) w_x + (A - \mu a) x w_{x-1} = \frac{c_1}{1 + x},$$

and w_x being hence found, we have

$$v_x = \left(\pi + \frac{B}{A}\right)\left(\pi + \frac{B}{A} - 1\right) \ldots (\pi + 1) w_x$$

for the complete integral.

2ndly. Let μ be determined so as if possible to cause the second term of (c) to vanish. This requires that we have

$$2a\mu + c = 0,$$

$$(b - a) \mu + e = 0,$$

and therefore imposes the condition

$$2ae + (b - a) c = 0.$$

Supposing this satisfied, we obtain, on making $\mu = \dfrac{-c}{2a}$,

$$v_x - \frac{h^2}{\pi \left(\pi + \dfrac{b}{a}\right)} \rho^2 v_x = \frac{1}{\mu^2 \pi \left(\pi + \dfrac{b}{a}\right)} x \mu^{-x+2} X,$$

or, representing any particular value of the second member by V,

$$v_x - \frac{h^2}{\pi \left(\pi + \dfrac{b}{a}\right)} \rho^2 v_x = V,$$

where

$$h = \frac{\sqrt{(c^2 - 4af)}}{c},$$

an equation which is integrable if $\dfrac{b}{a}$ be an odd number whether positive or negative. We must in such case assume

$$w_x - \frac{h^2}{\pi (\pi - 1)} \rho^2 w_x = W_x,$$

and determine first W_x and lastly v_x by A.

To found upon these results the conditions of solution of the general equation (a), viz.

$$(ax + b) u_x + (cx + e) u_{x-1} + (fx + g) u_{x-2} = X,$$

assume

$$fx + g = f (x' - 1),$$
$$u_x = t_{x'}.$$

Then

$$\left(ax' + b - a\frac{1+g}{f}\right) t_{x'}$$
$$+ \left(cx' + e - c\frac{1+f}{g}\right) t_{x'-1} + f(x'-1)\, t_{x'-2} = X',$$

comparing which with (b) we see that it is only necessary in the expression of the conditions already deduced to change

$$b \text{ into } b - \frac{a(1+g)}{f}, \quad e \text{ into } e - \frac{c(1+g)}{f}.$$

Solution of the above equation when $X = 0$ by definite integrals.

9.   If representing u_x by u we express (a) in the form

$$(ax + b) u + (cx + e) \, \epsilon^{-\frac{d}{dx}} u + (fx + g) \, \epsilon^{-2\frac{d}{dx}} u = 0,$$

or

$$x \left(a + c\epsilon^{-\frac{d}{dx}} + f\epsilon^{-2\frac{d}{dx}} \right) u + \left(b + e\epsilon^{-\frac{d}{dx}} + ge^{-2\frac{d}{dx}} \right) u = 0,$$

its solution in definite integrals may be obtained by Laplace's method for differential equations of the form

$$x\phi \left(\frac{d}{dx} \right) u + \psi \left(\frac{d}{dx} \right) u = 0,$$

each particular integral of which is of the form

$$u = C \int \frac{\epsilon^{xt + \int \frac{\psi(t)}{\phi(t)} dt}}{\phi(t)} \, dt,$$

the limits of the final integration being any roots of the equation

$$\epsilon^{xt + \int \frac{\psi(t)}{\phi(t)} dt} = 0.$$

See *Differential Equations*, Chap. XVIII.

The above solution is obtained by assuming $u = \int \epsilon^{xt} f(t) \, dt$, and then by substitution in the given equation and reduction obtaining a differential equation for determining the form of $f(t)$, and an algebraic equation for determining the limits. Laplace actually makes the assumption

$$u = \int t^x F(t) \, dt,$$

which differs from the above only in that $\log t$ takes the place of t and of course leads to equivalent results (*Théorie Analytique des Probabilités*, pp. 121, 135). And he employs this method with a view not so much to the solution of difficult equations as to the expression of solutions in forms convenient for calculation when functions of large numbers are involved.

Thus taking his first example, viz.

$$u_{x+1} - (x+1)\, u_x = 0,$$

and assuming $u_x = \int t^x F(t)\, dt$, we have

$$\int t^{x+1} F(t)\, dt - (x+1) \int t^x F(t)\, dt = 0\dots\dots\dots\dots(i).$$

But

$$(x+1) \int t^x F(t)\, dt = \int F(t)\,(x+1)\, t^x\, dt$$

$$= F(t)\, t^{x+1} - \int t^{x+1} F'(t)\, dt.$$

So that (i) becomes on substitution

$$\int t^{x+1} \{F(t) + F'(t)\}\, dt - F(t)\, t^{x+1} = 0,$$

and furnishes the two equations

$$F'(t) + F(t) = 0,$$

$$F(t)\, t^{x+1} = 0,$$

the first of which gives

$$F(t) = C\epsilon^{-t},$$

and thus reducing the second to the form

$$C\epsilon^{-t}\, t^{x+1} = 0,$$

gives for the limits $t = 0$ and $t = \infty$, on the assumption that $x+1$ is positive. Thus we have finally

$$u_x = C \int_0^\infty \epsilon^{-t} t^x\, dt,$$

the well known expression for $\Gamma(x+1)$. A peculiar method of integration is then applied to convert the above definite integral into a rapidly convergent series.

Discussion of the equation

$$(ax^2 + bx + c)\, u_x + (ex + f)\, u_{x-1} + gu_{x-2} = 0\dots\dots\dots(a).$$

10.   Let $u_x = \dfrac{\mu^x v_x}{1 \cdot 2 \dots x}$; then

$$\mu^2 (ax^2 + bx + c)\, v_x + \mu\,(ex + f)\, xv_{x-1} + gx\,(x-1)\, v_{x-2} = 0.$$

Whence, assuming $\pi = x \dfrac{\Delta}{\Delta x}$, $v = xD$, where $\Delta x = -1$, we have

$$\mu^2 (ax^2 + bx + c) \, v_x + \mu (ex + f) \, \rho v_x + g \rho^2 v_x = 0.$$

Therefore substituting $\pi + \rho$ for x, and developing by (13),

$$\mu^2 (a\pi^2 + b\pi + c) + \mu \{(2a\mu + e) \pi + (b - a) \mu + f\} \rho v_x$$
$$+ (\mu^2 a + \mu e + g) \rho^2 v_x = 0 \dots\dots\dots(b).$$

First, let μ be determined so as to satisfy the equation

$$a\mu^2 + e\mu + g = 0,$$

then

$$\mu (a\pi^2 + b\pi + c) \, v_x + \{(2a\mu + e) \pi + (b - a) \mu + f\} \rho v_x = 0.$$

Whence, by Art. 5,

$$v_x = \Sigma a_m \, x \, (x - 1) \, \dots \, (x - m + 1),$$

the successive values of m being determined by the equation

$$\mu^2 (am^2 + bm + c) \, a_m + \{(2a\mu^2 + e\mu) \, m + (b - a) \, \mu^2 + f\mu\} \, a_{m-1} = 0,$$

or $$a_m = - \frac{(2a\mu + e) \, m + (b - a) \, \mu + f}{\mu \, (am^2 + bm + c)} \, a_{m-1}.$$

Represent this equation in the form

$$a_m = - f(m) \, a_{m-1},$$

and let the roots of the equation

$$am^2 + bm + c = 0$$

be α and β, then,

$$v_x = C \{x^{(\alpha)} - f(\alpha + 1) \, x^{(\alpha+1)} + f(\alpha + 1) f(\alpha + 2) \, x^{(\alpha+2)} - \&\text{c.}\}$$
$$+ C' \{x^{(\beta)} - f(\beta+1) \, x^{(\beta+1)} + f(\beta+1) f(\beta+2) \, x^{(\beta+2)} - \&\text{c.}\} \dots (c),$$

where generally $x^{(p)} = x \, (x - 1) \, \dots \, (x - p + 1)$.

One of these series will terminate whenever the value of m given by the equation

$$(2a\mu + e)\, m + (b - a)\, \mu + f = 0,$$

exceeds by an integer either root of the equation

$$am^2 + bm + c = 0.$$

The solution may then be completed as in the last example.

Secondly, let μ be determined if possible so as to cause the second term of (b) to vanish. This gives

$$2a\mu + e = 0,$$
$$(b - a)\, \mu + f = 0,$$

whence, eliminating μ, we have the condition

$$2af + (a - b)\, e = 0.$$

This being satisfied, and μ being assumed equal to $-\dfrac{e}{2a}$, (b) becomes

$$(a\pi^2 + b\pi + c)\, v_x - \frac{a\,(e^2 - 4ag)}{e^2}\, \rho^2 v_x = 0.$$

Or putting $\qquad h = \dfrac{\sqrt{(e^2 - 4ag)}}{e},$

$$v_x - \frac{h^2}{\pi^2 + \dfrac{b}{a}\,\pi + \dfrac{c}{a}}\, \rho^2 v_x = 0,$$

and is integrable in finite terms if the roots of the equation

$$m^2 + \frac{b}{a}\, m + \frac{c}{a} = 0$$

differ by an odd number.

Discussion of the equation

$$(ax^2 + bx + c)\, \Delta^2 u_x + (ex + f)\, \Delta u_x + g u_x = 0.$$

11. By resolution of its coefficients this equation is reducible to the form

$$a\,(x - \alpha)\,(x - \beta)\, \Delta^2 u_x + e\,(x - \gamma)\, \Delta u_x + g u_x = 0 \;\dots\; (a).$$

Now let $x - a = x' + 1$ and $u_x = v_{x'}$, then we have

$$a (x' + 1) (x' + a - \beta + 1) \Delta^2 v_{x'}$$
$$+ e (x' + a - \gamma + 1) \Delta v_{x'} + g v_{x'} = 0,$$

or, dropping the accent,

$$a (x + 1) (x + a - \beta + 1) \Delta^2 v_x$$
$$+ e (x + a - \gamma + 1) \Delta v_x + g v_x = 0 \ldots (b).$$

If from the solution of this equation v_x be obtained, the value of u_x will thence be deduced by merely changing x into $x - a - 1$.

Now multiply (b) by x, and assume

$$\pi = x \frac{\Delta}{\Delta x} \quad \rho = xD,$$

where $\Delta x = 1$.   Then, since by (20),

$$x (x + 1) \Delta^2 v_x = \pi (\pi - 1) v_x,$$

we have

$$a (x + a - \beta + 1) \pi (\pi - 1) v_x$$
$$+ e (x + a - \gamma + 1) \pi v_x + g v_x = 0.$$

But $x = -\pi + \rho$, therefore substituting, and developing the coefficients we have on reduction

$$\pi \{a (\pi - a + \beta - 1) (\pi - 1) + e (\pi - a + \gamma - 1) + g\} v_x$$
$$- \{a (\pi - 1) (\pi - 2) + e (\pi - 1) + g\} \rho v_x = 0 \ldots (c).$$

And this is a binomial equation whose solutions in series are of the form

$$v_x = \Sigma a_m x (x + 1) \ldots (x + m - 1),$$

the lowest value of m being a root of the equation

$$m \{a (m - a + \beta - 1) (m - 1) + e (m - a + \gamma - 1) + g\} = 0 \ldots (d),$$

corresponding to which value a_m is an arbitrary constant, while all succeeding values of a_m are determined by the law

$$a_m = \frac{a (m - 1) (m - 2) + e (m - 1) + g}{m \{a (m - a + \beta - 1) (m - 1) + e (m - a + \gamma - 1) + g\}} a_{m-1}.$$

Hence the series terminates when a root of the equation

$$a\,(m-1)\,(m-2) + e\,(m-1) + g = 0 \ldots\ldots\ldots (e)$$

is equal to, or exceeds by an integer, a root of the equation (d).

As a particular root of the latter equation is 0, a particular finite solution may therefore always be obtained when (e) is satisfied either by a vanishing or by a positive integral value of m.

12.   The general theorem expressed by (38) admits of the following generalization, viz.

$$F\{\pi,\, \phi\,(\pi)\,\rho^{n}\} = P_{n}\frac{\phi\,(\pi)}{\psi\,(\pi)}\, F\{\pi,\, \psi\,(\pi)\,\rho^{n}\}\, P_{n}\frac{\psi\,(\pi)}{\phi\,(\pi)}.$$

The ground of this extension is that the symbol π, which is here newly introduced under F, combines with the same symbol π in the composition of the forms $P_{n}\dfrac{\phi\,(\pi)}{\psi\,(\pi)},\ P_{n}\dfrac{\psi\,(\pi)}{\phi\,(\pi)}$ external to F, as if π were algebraic.

And this enables us to transform some classes of equations which are not binomial. Thus the solution of the equation

$$f_{0}\,(\pi)\,u + f_{1}\,(\pi)\,\phi\,(\pi)\,\rho u + f_{2}\,(\pi)\,\phi\,(\pi)\,\phi\,(\pi-1)\,\rho^{2}u = U$$

will be made to depend upon that of the equation

$$f_{0}\,(\pi)\,v + f_{1}\,(\pi)\,\psi\,(\pi)\,\rho v + f_{2}\,(\pi)\,\psi\,(\pi)\,\psi\,(\pi-1)\,\rho^{2}v = P_{1}\frac{\psi\,(\pi)}{\phi\,(\pi)}\,U$$

by the assumption

$$u = P_{1}\frac{\phi\,(\pi)}{\psi\,(\pi)}\,v.$$

13.   While those transformations and reductions which depend upon the fundamental laws connecting π and ρ, and are expressed by (4), are common in their application to differential equations and to equations of differences, a marked difference exists between the two classes of equations as respects the conditions of finite solution. In differential equations where $\pi = \dfrac{d}{d\theta},\ \rho = \epsilon^{\theta}$, there appear to be three primary integrable forms for binomial equations, viz.

B. F. D.   12

$$u + \frac{a\pi + b}{c\pi + e}\rho^n u = U,$$

$$u + \frac{a(\pi - n)^2 + b}{\pi\left(\pi - \dfrac{n}{2}\right)}\rho^n u = U,$$

$$u + \frac{\left(\pi - \dfrac{n}{2}\right)(\pi - n)}{a\pi^2 + b}\rho^n u = U,$$

primary in the sense implied by the fact that every binomial equation, whatsoever its order, which admits of *finite* solution, is reducible to some one of the above forms by the transformations of Art. 7, founded upon the formal laws connecting π and ρ. In equations of differences but one primary integrable form for binomial equations is at present known, viz.

$$u + \frac{1}{a\pi + b}\rho u = U,$$

and this is but a particular case of the first of the above forms for differential equations. General considerations like these may serve to indicate the path of future inquiry.

EXERCISES.

1. Of what theorem in the differential calculus does (20), Art. 4, constitute a generalization?

2. Solve the equation

$$x(x+1)\Delta^2 u + x\Delta u - n^2 u = 0.$$

3. Solve by the methods of Art. 7 the equation of differences of Ex. 1, Art. 5, supposing a to be a positive odd number.

4. Solve by the same methods the same equation supposing a to be a negative odd number.

CHAPTER X.

OF EQUATIONS OF PARTIAL AND OF MIXED DIFFERENCES, AND OF SIMULTANEOUS EQUATIONS OF DIFFERENCES.

1. IF $u_{x,y}$ be any function of x and y, then

$$\left.\begin{aligned}\frac{\Delta}{\Delta x} u_{x,y} &= \frac{u_{x+\Delta x.y} - u_{x,y}}{\Delta x}, \\ \frac{\Delta}{\Delta y} u_{x\,y} &= \frac{u_{x,y+\Delta y} - u_{x,y}}{\Delta y}.\end{aligned}\right\} \quad \ldots\ldots\ldots\ldots (1).$$

These are, properly speaking, the coefficients of partial differences of the first order of $u_{x,y}$. But on the assumption that Δx and Δy are each equal to unity, an assumption which we can always legitimate, Chap. I. Art. 2, the above are the *partial differences* of the first order of $u_{x,y}$.

On the same assumption the general form of a partial difference of $u_{x,y}$ is

$$\frac{\Delta^{m+n}}{(\Delta x)^m (\Delta y)^n} u_{x,y}, \quad \text{or} \quad \left(\frac{\Delta}{\Delta x}\right)^m \left(\frac{\Delta}{\Delta y}\right)^n u_{x,y} \ldots\ldots\ldots (2).$$

When the form of $u_{x,y}$ is given, this expression is to be interpreted by performing the successive operations indicated, each elementary operation being of the kind indicated in (1).

Thus we shall find

$$\frac{\Delta^2}{\Delta x \Delta y} u_{x,y} = u_{x+2\Delta x, \Delta y} - 2u_{x+\Delta x, y+\Delta y} + u_{x, y+2\Delta y}.$$

It is evident that the operations $\dfrac{\Delta}{\Delta x}$ and $\dfrac{\Delta}{\Delta y}$ in combination are *commutative*.

Again, the symbolical expression of $\dfrac{\Delta}{\Delta x}$ in terms of $\dfrac{d}{dx}$ being

$$\frac{\Delta}{\Delta x} = \frac{\epsilon^{\Delta x \frac{d}{dx}} - 1}{\Delta x} \quad \dots\dots\dots\dots\dots \text{(3)},$$

in which Δx is an absolute constant, it follows that

$$\left(\frac{\Delta}{\Delta x}\right)^n = \frac{\epsilon^{n\Delta x \frac{d}{dx}} - n\epsilon^{(n-1)\Delta x \frac{d}{dx}} + \dfrac{n\,(n-1)}{1\,.\,2}\,\epsilon^{(n-2)\Delta x \frac{d}{dx}} - \&\text{c.}}{(\Delta x)^n},$$

and therefore

$$\left(\frac{\Delta}{\Delta x}\right)^n u_{x,y} = \left\{ u_{x+n\Delta x,\,y} - n u_{x+(n-1)\Delta x,\,y} \right.$$

$$\left. + \frac{n\,(n-1)}{1\,.\,2}\, u_{x+(n-2)\Delta x,\,y} - \&\text{c.} \right\} \div (\Delta x)^n \dots\dots \text{(4)}.$$

So, also, to express $\left(\dfrac{\Delta}{\Delta x}\right)^m \left(\dfrac{\Delta}{\Delta y}\right)^n u_{x,y}$ it would be necessary to substitute for $\dfrac{\Delta}{\Delta x}$, $\dfrac{\Delta}{\Delta y}$ their symbolical expressions, to effect their symbolical expansions by the binomial theorem, and then to perform the final operations on the subject function $u_{x,y}$.

Though in what follows each increment of an independent variable will be supposed equal to unity, it will still be necessary to retain the notation $\dfrac{\Delta}{\Delta x}$, $\dfrac{\Delta}{\Delta y}$ for the sake of distinction, or to substitute some notation equivalent by definition, e. g. Δ_x, Δ_y.

These things premised, we may define an equation of partial differences as an equation expressing an algebraic relation between any partial differences of a function $u_{x,y,z...}$, the function itself, and the independent variables $x, y, z \dots$ Or instead of the partial *differences* of the dependent function, its successive *values* corresponding to successive states of increment of the independent variables may be involved.

Thus
$$x \frac{\Delta}{\Delta x} u_{x,y} + y \frac{\Delta}{\Delta y} u_{x,y} = 0,$$

and
$$x u_{x+1,y} + y u_{x,y+1} - (x + y) u_{x,y} = 0,$$

are, on the hypothesis of Δx and Δy being each equal to unity, different but equivalent forms of the same equation of partial differences.

Equations of mixed differences are those in which the subject function is presented as modified both by operations of the form $\frac{\Delta}{\Delta x}$, $\frac{\Delta}{\Delta y}$, and by operations of the form $\frac{d}{dx}$, $\frac{d}{dy}$, singly or in succession. Thus

$$x \frac{\Delta}{\Delta x} u_{x,y} + y \frac{d}{dy} u_{x,y} = 0$$

is an equation of mixed differences. Upon the obvious subordinate distinction of equations of ordinary mixed differences and equations of partial mixed differences it is unnecessary to enter.

Before engaging in the discussion of any of the above classes of equations a remark must be made upon a question of analogy and of order.

The theory of partial differential equations is essentially dependent upon that of simultaneous differential equations of the ordinary kind, and therefore follows it. But it may be affirmed that in the present state of the Calculus of Finite Differences any such order of dependence between equations of partial differences and simultaneous equations of differences is of a quite subordinate importance. In nearly all the cases in which the finite solution of equations of partial differences is as yet possible it will be found to depend, directly or virtually, upon a symbolical reduction by which the proposed equation is presented as an ordinary equation of differences, the chief, or only, difficulty of the solution consisting in a difficulty of interpretation. And the same observation is equally applicable to equations of mixed differences, no general theory determining *a priori* the nature of the solution of equations of either species, partial or mixed, even when but of the first order, as yet existing. At the same time whatever belongs to the solution of simultaneous equations of differences—and specially whatever relates to the process of their reduction to *single* equations—is of common application, and of equivalent import, whether those equations are of the ordinary, of the partial, or of the mixed species. We therefore think it right to treat of the subjects of this chapter in the order in which they have been enumerated, i. e. we shall treat first of equations of partial differences; secondly, of equations of mixed differences; lastly, of simultaneous equations of all species.

Equations of Partial Differences.

2. When there are two independent variables x and y, while the coefficients are constant and the second member is 0, the proposed equation may be presented, according to convenience, in any of the forms

$$F(\Delta_x, \Delta_y)\, u = 0, \qquad F(D_x, D_y)\, u = 0,$$

$$F(\Delta_x, D_y)\, u = 0, \qquad F(D_x, \Delta_y)\, u = 0.$$

Now the symbol of operation relating to x, viz. Δ_x or D_x, combines with that relating to y, viz. Δ_y or D_y, as a constant with a constant. Hence a symbolical solution will be obtained by replacing one of the symbols by a constant quantity a, integrating the *ordinary* equation of differences which results, replacing a by the symbol in whose place it stands, and the arbitrary constant by an arbitrary function of the independent variable to which that symbol has reference. This arbitrary function must *follow* the expression which contains the symbol corresponding to a.

The condition last mentioned is founded upon the interpretation of $(D-a)^{-i}X$, upon which the solution of ordinary equations of differences with constant coefficients is ultimately dependent. For (Chap. VII. Art. 6)

$$(D-a)^{-i}X = a^{x-i}\Sigma^i a^{-x}X,$$

whence

$$(D-a)^{-i}0 = a^{x-i}\Sigma^i 0$$
$$= a^{x-i}(c_0 + c_1 x \ldots + c_{n-1} x^{n-1}),$$

the constants *following* the factor involving a.

The difficulty of the solution is thus reduced to the difficulty of interpreting the symbolical result.

Ex. 1. Thus the solution of the equation $u_{x+1} - au_x = 0$, of which the symbolical form is

$$D_x u_x - au_x = 0,$$

being

$$u_x = Ca^x,$$

the solution of the equation $u_{x+1, y} - u_{x, y+1} = 0$, of which the symbolic form is

$$D_x u_{x, y} - D_y u_{x, y} = 0,$$

will be

$$u_{x, y} = (D_y)^x \phi(y).$$

To interpret this we observe that since $D_y = \epsilon^{\frac{d}{dy}}$ we have

$$u_{x, y} = \epsilon^{x \frac{d}{dy}} \phi(y)$$
$$= \phi(y + x).$$

Ex. 2. Given $u_{x+1, y+1} - u_{x, y+1} - u_{x, y} = 0$.

This equation, on putting u for $u_{x, y}$, may be presented in the form

$$D_y \Delta_x u - u = 0 \ldots\ldots\ldots\ldots\ldots\ldots(1).$$

Now replacing D_y by a, the solution of the equation

$$a\Delta_x u - u = 0$$

is

$$u = (1 + a^{-1})^x C,$$

therefore the solution of (1) is

$$u = (1 + D_y^{-1})^x \phi(y) \ldots\ldots\ldots\ldots(2),$$

where $\phi(y)$ is an arbitrary function of y. Now, developing the binomial, and applying the theorem

$$D_y^{-n}\phi(y) = \phi(y-n),$$

we find

$$u = \phi(y) + x\phi(y-1) + \frac{x(x-1)}{1.2}\phi(y-2) + \&c.... \ (3),$$

which is finite when x is an integer.

Or, expressing (2) in the form

$$u = (D_y+1)^x D_y^{-x}\phi(y),$$

developing the binomial in ascending powers of D_y, and interpreting, we have

$$u = \phi(y-x) + x\phi(y-x+1)$$

$$+ \frac{x(x-1)}{1.2}\phi(y-x+2) + \&c............(4).$$

Or, treating the given equation as an ordinary equation of differences in which y is the independent variable, we find as the solution

$$u = (\Delta_x)^{-y}\phi(x)..................... \ (5).$$

Any of these three forms may be used according to the requirements of the problem.

Thus if it were required that when $x = 0$, u should assume the form ϵ^{my}, it would be best to employ (3) or to revert to (2) which gives $\phi(y) = \epsilon^{my}$, whence

$$u = (1 + D_y^{-1})^x \epsilon^{my}$$

$$= (1 + \epsilon^{-\frac{d}{dy}})^x \epsilon^{my}$$

$$= (1 + \epsilon^{-m})^x \epsilon^{my}..................... \ (6).$$

3.   There is another method of integrating this class of equations with constant coefficients which deserves attention. We shall illustrate it by the last example.

Assume $u_{x,y} = \Sigma\, Ca^x b^y$, then substituting in the given equation we find as the sole condition

$$ab - b - 1 = 0.$$

Hence

$$a = \frac{1+b}{b},$$

and substituting,

$$u_{x,y} = \Sigma\, C\, (1+b)^x b^{y-x}.$$

As the summation denoted by Σ has reference to all possible values of b, and C may vary in a perfectly arbitrary manner for different values of b, we shall best express the character of the solution by making C an arbitrary function of b and changing the summation into an integration extended from $-\infty$ to ∞. Thus we have

$$u_{x,y} = \int_{-\infty}^{\infty} b^{y-x}(1+b)^x \phi(b)\, db.$$

As $\phi(b)$ may be discontinuous, we may practically make the limits of integration what we please by supposing $\phi(b)$ to vanish when these limits are exceeded.

If we develope the binomial in ascending powers of b, we have

$$u_{x,y} = \int_{-\infty}^{\infty} b^{y-x} \phi(b)\, db + x \int_{-\infty}^{\infty} b^{y-x+1} \phi(b)\, db$$

$$+ \frac{x(x-1)}{1.2} \int_{-\infty}^{\infty} b^{y-x+2} \phi(b)\, db + \&c. \ldots\ldots (7).$$

Now

$$\int_{-\infty}^{\infty} b^{\theta} \phi(b)\, db = \psi(\theta),$$

$\psi(\theta)$ being arbitrary if $\phi(b)$ is; hence

$$u_{x,y} = \psi(y-x) + x\psi(y-x+1) + \frac{x(x-1)}{1.2}\, \psi(y-x+2) + \&c.,$$

which agrees with (4).

Although it is usually much the more convenient course to employ the symbolical method of Art. 2, yet cases may arise in which the expression of the solution by means of a definite integral will be attended with advantage; and the connexion of the methods is at least interesting.

Ex. 3.   Given $\Delta^2_x u_{x-1,y} = \Delta^2_y u_{x,y-1}$.

Replacing $u_{x,y}$ by u, we have

$$(\Delta^2_x D_x^{-1} - \Delta^2_y D_y^{-1})\, u = 0,$$

or $$(\Delta^2_x D_y - \Delta^2_y D_x)\, u = 0.$$

But $$\Delta_x = D_x - 1, \quad \Delta_y = D_y - 1 \; ;$$

therefore $$(D_x^2 D_y + D_y - D_y^2 D_x - D_x)\, u = 0,$$

or $$(D_x D_y - 1)\,(D_x - D_y)\, u = 0.$$

This is resolvable into the two equations

$$(D_x D_y - 1)\, u = 0, \quad (D_x - D_y)\, u = 0.$$

The first gives

$$D_x u - D_y^{-1} u = 0,$$

of which the solution is

$$u = (D_y^{-1})^x\, \phi\,(y)$$
$$= \phi\,(y - x).$$

The second gives, by Ex. 1,

$$u = \psi\,(x + y).$$

Hence the complete integral is

$$u = \phi\,(y - x) + \psi\,(y + x).$$

4.   Upon the result of this example an argument has been founded for the discontinuity of the arbitrary functions which occur in the solution of the partial differential equation

$$\frac{d^2 u}{dx^2} - \frac{d^2 u}{dy^2} = 0,$$

and thence, by obvious transformation, in that of the equation

$$\frac{d^2u}{dx^2} - a^2 \frac{d^2u}{dt^2} = 0.$$

It is perhaps needless for me, after what has been said in Chap. VIII. Arts. 4 and 5, to add that I regard the *argument* as unsound. Analytically such questions depend upon the following, viz. whether in the proper sense of the term limit, we can regard $\sin x$ and $\cos x$ as tending to the limit 0, when x tends to become infinite.

5.   When together with Δ_x and Δ_y one only of the independent variables, e.g. x, is involved, or when the equation contains both the independent variables, but only one of the operative symbols Δ_x, Δ_y, the same principle of solution is applicable. A symbolic solution of the equation

$$F(x, \Delta_x, \Delta_y)\, u = 0$$

will be found by substituting Δ_y for a and converting the arbitrary constant into an arbitrary function of y in the solution of the ordinary equation

$$F(x, \Delta_x, a)\, u = 0.$$

And a solution of the equation

$$F(x, y, \Delta_x) = 0$$

will be obtained by integrating as if y were a constant, and replacing the arbitrary constant, as before, by an arbitrary function of y. But if x, y, Δ_x and Δ_y are involved together, this principle is no longer applicable. For although y and Δ_y are constant relatively to x and Δ_x, they are not so with respect to each other. In such cases we must endeavour by a change of variables, or by some tentative hypothesis as to the form of the solution, to reduce the problem to easier conditions.

The extension of the method to the case in which the second member is not equal to 0 involves no difficulty.

Ex. 4.   Given $u_{x,y} - x u_{x-1,y-1} = 0.$

Writing u for $u_{x,y}$ the equation may be expressed in the form

$$u - xD_x^{-1}D_y^{-1}u = 0 \dots\dots\dots\dots\dots(1).$$

Now replacing D_y^{-1} by a, the solution of

$$u - axD_x^{-1}u = 0 \text{ or } u_x - axu_{x-1} = 0$$

is $$Cx(x-1)\dots 1 \cdot a^x.$$

Wherefore, changing a into D_y^{-1}, the solution of (1) is

$$u = (D_y^{-1})^x \, x\,(x-1)\dots 1 \cdot \phi(y)$$
$$= x\,(x-1)\dots 1 \cdot (D_y^{-x})\,\phi(y)$$
$$= x\,(x-1)\dots 1 \cdot \phi(y-x).$$

6.   Laplace has shewn how to solve any linear equation in the successive terms of which the progression of differences is the same with respect to one independent variable as with respect to the other.

The given equation being

$$A_{x,y}u_{x,y} + B_{x,y}u_{x-1,y-1} + C_{x,y}u_{x-2,y-2} + \&c. = V_{x,y},$$

$A_{x,y}$, $B_{x,y}$, &c., being functions of x and y, let $y = x - k$; then substituting and representing $u_{x,y}$ by v_x, the equation assumes the form

$$X_0 v_x + X_1 v_{x-1} + X_2 v_{x-2} + \&c. = X,$$

X_0, X_1 ... X being functions of x. This being integrated, k is replaced by $x - y$, and the arbitrary constants by arbitrary functions of $x - y$.

The ground of this method is that the progression of differences in the given equation is such as to leave $x - y$ unaffected, for when x and y change by equal differences $x - y$ is unchanged. Hence if $x - y$ is represented by k and we take x and k for the new variables, the differences now having reference to x only, we can integrate as if k were constant.

Applying this method to the last example, we have

$$v_x - xv_{x-1} = 0,$$

$$v_x = cx\,(x-1)\,\ldots\,1,$$

$$u_{x,y} = x\,(x-1)\,\ldots\,1\,.\,\phi\,(x-y),$$

which agrees with the previous result.

The method may be generalized. Should any linear function of x and y, e.g. $x+y$, be invariable, we may by assuming it as one of the independent variables, so to speak reduce the equation to an ordinary equation of differences; but arbitrary functions of the element in question must take the place of arbitrary constants.

Ex. 5.   Given $u_{x,y} - pu_{x+1,y-1} - (1-p)\,u_{x-1,y+1} = 0$.

Here $x+y$ is invariable. Now the integral of

$$v_x - pv_{x+1} - (1-p)\,v_{x-1} = 0$$

is
$$v_x = c + c'\left(\frac{1-p}{p}\right)^x.$$

Hence, that of the given equation is

$$u_{x,y} = \phi\,(x+y) + \left(\frac{1-p}{p}\right)^x \psi\,(x+y).$$

7. Equations of partial differences are of frequent occurrence in the theory of games of chance. The following is an example of the kind of problems in which they present themselves.

Ex. 6. A and B engage in a game, each step of which consists in one of them winning a counter from the other. At the commencement, A has x counters and B has y counters, and in each successive step the probability of A's winning a counter from B is p, and therefore of B's winning a counter from A, $1-p$. The game is to terminate when either of the two has n counters. What is the probability of A's winning it ?

Let $u_{x,y}$ be the probability that A will win it, any positive values being assigned to x and y.

Now A's winning the game may be resolved into two alternatives, viz. 1st, His winning the first step, and afterwards winning the game. 2ndly, His losing the first step, and afterwards winning the game.

The probability of the first alternative is $pu_{x+1,\,y-1}$, for after A's winning the first step, the probability of which is p, he will have $x+1$ counters, B, $y-1$ counters, therefore the probability that A will then win is $u_{x+1,y-1}$. Hence the probability of the combination is $pu_{x+1,y-1}$.

The probability of the second alternative is in like manner $(1-p)\,u_{x-1,y+1}$.

Hence, the probability of any event being the sum of the probabilities of the alternatives of which it is composed, we have as the equation of the problem

$$u_{x,y} = pu_{x+1\ y-1} + (1-p)\,u_{x-1,y+1} \cdots\cdots\cdots (1),$$

the solution of which is, by the last example,

$$u_{x,y} = \phi\,(x+y) + \left(\frac{1-p}{p}\right)^{x} \psi\,(x+y).$$

It remains to determine the arbitrary functions.

The number of counters $x+y$ is invariable through the game. Represent it by m, then

$$u_{x,y} = \phi\,(m) + \left(\frac{1-p}{p}\right)^{x} \psi\,(m).$$

Now A's success is certain if he should ever be in possession of n counters. Hence, if $x=n$, $u_{x,y}=1$. Therefore

$$1 = \phi\,(m) + \left(\frac{1-p}{p}\right)^{n} \psi\,(m).$$

Again, A loses the game if ever he have only $m-n$ counters, since then B will have n counters. Hence

$$0 = \phi\,(m) + \left(\frac{1-p}{p}\right)^{m-n} \psi\,(m).$$

The last two equations give, on putting $P = \dfrac{1-p}{p}$,

$$\phi(m) = \frac{-P^{m-n}}{P^n - P^{m-n}}, \psi(m) = \frac{1}{P^n - P^{m-n}},$$

whence

$$u_{x,y} = \frac{P^{n-y} - 1}{P^{2n-x-y} - 1}$$

$$= \frac{p^{n-y} - (1-p)^{n-y}}{p^{2n-x-y} - (1-p)^{2n-x-y}} \cdots\cdots\cdots (2),$$

which is the probability that A will win the game.

Symmetry therefore shews that the probability that B will win the game is

$$\frac{(1-p)^{n-x} - p^{n-x}}{(1-p)^{2n-x-y} - p^{2n-x-y}} \cdots\cdots\cdots\cdots (3),$$

and the sum of these values will be found to be unity.

The problem of the 'duration of play' in which it is proposed to find the probability that the game conditioned as above will terminate at a particular step, suppose the rth, depends on the same equation of partial differences, but it involves great difficulty. A very complete solution, rich in its analytical consequences, will be found in a memoir by the late Mr Leslie Ellis (*Cambridge Mathematical Journal*, Vol. IV. p. 182).

Method of Generating Functions.

8. Laplace usually solves problems of the above class by the method of generating functions, the most complete statement of which is contained in the following theorem.

Let u be the generating function of $u_{m,n...}$, so that

$$u = \Sigma u_{m,n...} \, x^m y^n ...,$$

then making $x = \epsilon^\theta$, $y = \epsilon^{\theta'}$ &c., we have

$$S\phi\left(\frac{d}{d\theta}, \frac{d}{d\theta'}, \cdots\right) \epsilon^{\theta + g\theta'\cdots} u$$

$$= \Sigma \left\{ S\phi(m, n ...) u_{m-p, n-q...} \right\} \epsilon^{m\theta + n\theta'\cdots} \cdots\cdots\cdots (1).$$

Here, while Σ denotes summation with respect to the terms of the development of u, S denotes summation with respect to the operations which would constitute the first member a member of a linear differential equation, and the bracketed portion of the second member a member of an equation of differences.

Hence it follows that if we have a linear equation of differences of the form

$$S\phi\,(m, n \dots)\, u_{m-p, n-q\dots} = 0 \dots\dots\dots\dots\dots (2),$$

the equation (1) would give for the general determination of the generating function u the linear differential equation

$$\Sigma\phi\left(\frac{d}{d\theta},\ \frac{d}{d\theta'} \dots\right)\epsilon^{p\theta+q\theta'\dots}u = 0 \dots\dots\dots\dots (3).$$

But if there be given certain initial values of $u_{m, n}$ which the equation of differences does not determine, then, corresponding to such initial values, terms will arise in the second member of (1) so that the differential equation will assume the form

$$S\phi\left(\frac{d}{d\theta},\ \frac{d}{d\theta'} \dots\right)\epsilon^{p\theta+q\theta'\dots}u = F(m, n\dots)\dots\dots\dots(4).$$

If the equation of differences have constant coefficients the differential equation merges into an algebraic one, and the generating function will be a rational fraction. This is the case in most, if not all, of Laplace's examples.

It must be borne in mind that the discovery of the generating function is but a step toward the solution of the equation of differences, and that the next step, viz. the discovery of the general term of its development by some *independent* process, is usually far more difficult than the direct solution of the original equation of differences would be. As I think that in the present state of analysis the interest which belongs to this application of generating functions is chiefly historical, I refrain from adding examples.

Equations of mixed differences.

9. When an 'equation of mixed differences admits of re-solution into an equation of simple differences and a differen-tial equation, the process of solution is obvious.

Ex. 7. Thus the equation

$$\Delta \frac{du}{dx} - a\Delta u - b\cdot\frac{du}{dx} + abu = 0$$

being presented in the form

$$\left(\frac{d}{dx} - a\right)(\Delta - b)\, u = 0,$$

the complete value of u will evidently be the sum of the values given by the resolved equations

$$\frac{du}{dx} - au = 0, \quad \Delta u - bu = 0.$$

Hence

$$u = c_1 \epsilon^{ax} + c_2 (1 + b)^x,$$

where c_1 is an absolute, c_2 a periodical constant.

Ex. 8. Again, the equation $\Delta y = x\dfrac{d}{dx}\Delta y + \left(\dfrac{d}{dx}\Delta y\right)^2$ being resolvable into the two equations,

$$\Delta y = z, \quad z = x\frac{dz}{dx} + \left(\frac{dz}{dx}\right)^2,$$

we have, on integration,

$$z = cx + c^2,$$

$$y = \Sigma z = \frac{cx\,(x-1)}{2} + c^2 x + C,$$

where c is an absolute, and C a periodical constant.

Equations of mixed differences are reducible to differential equations of an exponential form by substituting for D_x or Δ_x their differential expressions $\epsilon^{\frac{d}{dx}}$, $\epsilon^{\frac{d}{dx}} - 1$.

Ex. 9.   Thus the equation $\Delta u - \dfrac{du}{dy} = 0$ becomes

$$\left(\epsilon^{\frac{d}{dy}} - 1 - \frac{d}{dy} \right) u = 0,$$

and its solution will therefore be

$$u = \Sigma c_m \epsilon^{my},$$

the values of m being the different roots of the equation

$$\epsilon^m - 1 - m = 0.$$

10.   Laplace's method for the solution of a class of partial differential equations (*Diff. Equations*, p. 440) has been extended by Poisson to the solution of equations of mixed differences of the form

$$\frac{du_{x+1}}{dx} + L \frac{du_x}{dx} + M u_{x+1} + N u_x = V \ \ldots\ldots\ldots(1),$$

where L, M, N, V are functions of x.

Writing u for u_x, and expressing the above equation in the form

$$\frac{d}{dx} Du + L \frac{d}{dx} u + MDu + Nu = V,$$

it is easily shewn that it is reducible to the form

$$\left(\frac{d}{dx} + M \right)(D + L)u + (N - LM - L')u = V,$$

where $L' = \dfrac{dL}{dx}$.   Hence if we have

$$N - LM - L' = 0 \ldots\ldots\ldots\ldots\ldots(2),$$

the equation becomes

$$\left(\frac{d}{dx}+M\right)(D+L)\,u=V,$$

which is resolvable by the last section into an equation of mixed differences and a differential equation.

But if the above condition be not satisfied, then, assuming

$$(D+L)\,u=v\ldots\ldots\ldots\ldots\ldots(3),$$

we have

$$\left(\frac{d}{dx}+M\right)v+(N-LM-L')\,u=V,$$

whence

$$u=\frac{-\left(\frac{d}{dx}+M\right)v+V}{N-LM-L'}\ldots\ldots\ldots\ldots(4),$$

which is expressible in the form

$$u=A_x\frac{dv}{dx}+B_xv+C_x.$$

Substituting this value in (3) we have

$$A_{x+1}\frac{d}{dx}Dv+LA_x\frac{dv}{dx}+B_{x+1}Dv$$

$$+(LB_x-1)\,v_x=-\,C_{x+1}-LC_x,$$

which, on division by A_{x+1}, is of the form

$$\frac{d}{dx}Dv+L_1\frac{dv}{dx}+M_1Dv+N_1v=V_1.$$

The original form of the equation is thus reproduced with altered coefficients, and the equation is resolvable as before into an equation of mixed differences and a differential equation, if the condition

$$N_1-L_1M_1-L_1'=0\ldots\ldots\ldots\ldots(5)$$

is satisfied. If not, the operation is to be repeated.

13—2

An inversion of the order in which the symbols $\dfrac{d}{dx}$ and D are employed in the above process leads to another reduction similar in its general character.

Presenting the equation in the form

$$(D+L)\left(\frac{d}{dx}+M_{-1}\right)u+(N-LM_{-1})\,u=V$$

where $M_{-1}=D^{-1}M$, its direct resolution into an equation of mixed differences and a differential equation is seen to involve the condition

$$N-LM_{-1}=0\ldots\ldots\ldots\ldots\ldots\ldots(6).$$

If this equation be not satisfied, assume

$$\left(\frac{d}{dx}+M_{-1}\right)u=v,$$

and proceeding as before a new equation similar in form to the original one will be obtained to which a similar test, or, that test failing, a similar reduction may again be applied.

Ex. 10.   Given $\dfrac{du_{x+1}}{dx}-a\,\dfrac{du_x}{dx}+(x\pm n)\,u_{x+1}-axu_x=0.$

This is the most general of Poisson's examples. Taking first the lower sign we have

$$L=-a,\ \ M=x-n,\ \ N=-ax.$$

Hence the condition (2) is not satisfied. But (3) and (4) give

$$(D-a)\,u=v,$$

$$u=\frac{\dfrac{dv}{dx}+(x-n)\,v}{an}\,,$$

whence

$$(D-a)\left\{\frac{\dfrac{dv}{dx}+(x-n)\,v}{an}\right\}=v,$$

or, on reducing,

$$\frac{dv_{x+1}}{dx} - a\frac{dv_x}{dx} + \{x - (n-1)\}\, v_{x+1} - axv_x = 0.$$

Comparing this with the given equation, we see that n reductions similar to the above will result in an equation of the form

$$\frac{dw_{x+1}}{dx} - a\frac{dw_x}{dx} + xw_{x+1} - axw_x = 0,$$

which, being presented in the form

$$\left(\frac{d}{dx} + x\right)(D - a)\, w_x = 0,$$

is resolvable into two equations of the unmixed character.

Poisson's second reduction applies when the upper sign is taken in the equation given; and thus the equation is seen to be integrable whenever n is an integer positive or negative.

Its actual solution deduced by another method will be given in the following section.

11.   Equations of mixed differences in whose coefficients x is involved only in the first degree admit of a symbolical solution founded upon the theorem

$$\left\{x + \phi'\left(\frac{d}{dx}\right)\right\}^{-1} X = \epsilon^{\phi\left(\frac{d}{dx}\right)} x^{-1} \epsilon^{-\phi\left(\frac{d}{dx}\right)} X \ldots\ldots\ldots\ldots (1).$$

(*Differential Equations*, p. 445.)

The following is the simplest proof of the above theorem. Since

$$\psi\left(\frac{d}{dx}\right) xu = \psi\left(\frac{d}{dx} + \frac{d'}{dx}\right) xu,$$

if in the second member $\dfrac{d'}{dx}$ operate on x only, and $\dfrac{d}{dx}$ on u, we have, on developing and effecting the differentiations which have reference to x,

$$\psi\left(\frac{d}{dx}\right)xu = x\psi\left(\frac{d}{dx}\right)u + \psi'\left(\frac{d}{dx}\right)u.$$

Let $\psi\left(\frac{d}{dx}\right)u = v$, then

$$\psi\left(\frac{d}{dx}\right)x\left\{\psi\left(\frac{d}{dx}\right)\right\}^{-1}v = \left\{x + \frac{\psi'\left(\frac{d}{dx}\right)}{\psi\left(\frac{d}{dx}\right)}\right\}v,$$

or if $\psi\left(\frac{d}{dx}\right)$ be replaced by $\epsilon^{\phi\left(\frac{d}{dx}\right)}$,

$$\epsilon^{\phi\left(\frac{d}{dx}\right)}xe^{-\phi\left(\frac{d}{dx}\right)}v = \left\{x + \phi'\left(\frac{d}{dx}\right)\right\}v.$$

Inverting the operations on both sides, which involves the inverting of the order as well as of the character of successive operations, we have

$$\left\{x + \phi'\left(\frac{d}{dx}\right)\right\}^{-1}v = \epsilon^{\phi\left(\frac{d}{dx}\right)}x^{-1}\epsilon^{-\phi\left(\frac{d}{dx}\right)}v,$$

the theorem in question.

Let us resume Ex. 10, which we shall express in the form

$$\frac{du_{x+1}}{dx} - a\,\frac{du_{x+1}}{dx} + (x+n)\,u_{x+1} - axu_x = 0 \dots\dots (a),$$

n being either positive or negative. Now putting u for u_x

$$\left\{\frac{d}{dx}\left(\epsilon^{\frac{d}{dx}} - a\right) + n\epsilon^{\frac{d}{dx}}\right\}u + x\left(\epsilon^{\frac{d}{dx}} - a\right)u = 0.$$

Let $\qquad\qquad \left(\epsilon^{\frac{d}{dx}} - a\right)u = z,$

then we have

$$\left\{\frac{d}{dx} + n\epsilon^{\frac{d}{dx}}\left(\epsilon^{\frac{d}{dx}} - a\right)^{-1}\right\}z + xz = 0.$$

Or,

$$\left(x + \frac{d}{dx} + \frac{n\epsilon^{\frac{d}{dx}}}{\epsilon^{\frac{d}{dx}} - a}\right) z = 0.$$

Hence,

$$z = \left(x + \frac{d}{dx} + \frac{n\epsilon^{\frac{d}{dx}}}{\epsilon^{\frac{d}{dx}} - a}\right)^{-1} 0,$$

and therefore by (1),

$$z = \epsilon^{\frac{1}{2}\left(\frac{d}{dx}\right)^2 + n\log\,(\epsilon^{\frac{d}{dx}} - a)}\, x^{-1}\, \epsilon^{-\frac{1}{2}\left(\frac{d}{dx}\right)^2 - n\log\,(\epsilon^{\frac{d}{dx}} - a)}\, 0$$

$$= (\epsilon^{\frac{d}{dx}} - a)^n\, \epsilon^{\frac{1}{2}\left(\frac{d}{dx}\right)^2}\, x^{-1}\, \epsilon^{-\frac{1}{2}\left(\frac{d}{dx}\right)^2}\, (\epsilon^{\frac{d}{dx}} - a)^{-n}\, 0 \dots (b).$$

It is desirable to transform a part of this expression. By (1), we have

$$\left(x + \frac{d}{dx}\right)^{-1} = \epsilon^{\frac{1}{2}\left(\frac{d}{dx}\right)^2}\, x^{-1}\, \epsilon^{-\frac{1}{2}\left(\frac{d}{dx}\right)^2},$$

and by another known theorem,

$$\left(\frac{d}{dx} + x\right)^{-1} = \epsilon^{-\frac{1}{2}x^2} \left(\frac{d}{dx}\right)^{-1} \epsilon^{\frac{1}{2}x^2}.$$

The right-hand members of these equations being sympolically equivalent, we may therefore give to (b) the form

$$z = (\epsilon^{\frac{d}{dx}} - a)^n\, \epsilon^{-\frac{x^2}{2}} \left(\frac{d}{dx}\right)^{-1} \epsilon^{\frac{x^2}{2}} (\epsilon^{\frac{d}{dx}} - a)^{-n}\, 0 \dots (c).$$

Now $u = (\epsilon^{\frac{d}{dx}} - a)^{-1} z$, therefore substituting, and replacing $\epsilon^{\frac{d}{dx}}$ by D,

$$u = (D-a)^{n-1}\, \epsilon^{-\frac{x^2}{2}} \left(\frac{d}{dx}\right)^{-1} \epsilon^{\frac{x^2}{2}} (D-a)^{-n}\, 0 \dots (A).$$

Two cases here present themselves.

First, let n be a positive integer; then since

$$(D-a)^{-n}\, 0 = a^x\, (c_0 + c_1 x \ldots + c_{n-1} x^{n-1}),$$

$$(D-a)^{n-1} = (\Delta + 1 - a)^{n-1},$$

we have

$$u = (\Delta + 1 - a)^{n-1}\, \epsilon^{-\frac{x^2}{2}} \left\{ C + \int \epsilon^{\frac{x^2}{2}}\, a^x\, (c_0 + c_1 x \ldots + c_{n-1} x^{n-1})\, dx \right\}$$

$$\ldots\ldots\ldots (d),$$

as the solution required.

This solution involves superfluous constants. For integrating by parts, we have

$$\int \epsilon^{\frac{x^2}{2}}\, a^x x^r\, dx = \epsilon^{\frac{x^2}{2}}\, a^x x^{r-1} + \log a \int \epsilon^{\frac{x^2}{2}}\, a^x x^{r-1}\, dx + (r-1) \int \epsilon^{\frac{x^2}{2}}\, a^x x^{r-2}\, dx,$$

and in particular when $r = 1$,

$$\int \epsilon^{\frac{x^2}{2}}\, a^x x\, dx = \epsilon^{\frac{x^2}{2}}\, a^x + \log a \int \epsilon^{\frac{x^2}{2}}\, a^x\, dx.$$

These theorems enable us, r being a positive integer, to reduce the above general integral to a linear function of the elementary integrals $\int \epsilon^{\frac{x^2}{2}}\, a^x\, dx$, and of certain algebraic terms of the form $\epsilon^{\frac{x^2}{2}}\, a^x x^m$, where m is an integer less than r.

Now if we thus reduce the integrals involved in (d), it will be found that the algebraic terms vanish.

For

$$(\Delta + 1 - a)^{n-1}\, \epsilon^{-\frac{x^2}{2}}\, (\epsilon^{\frac{x^2}{2}}\, a^x x^m) = (\Delta + 1 - a)^{n-1}\, a^x x^m$$

$$= a^{x+n-1}\, \Delta^{n-1} x^m$$

$$= 0,$$

since m is less than r, and the greatest value of r is $n-1$.

It results therefore that (d) assumes the simpler form,

$$u = (\Delta + 1 - a)^{n-1}\, \epsilon^{-\frac{x^2}{2}}\, \left(C_0 + C_1 \int \epsilon^{\frac{x^2}{2}}\, a^x\, dx \right);$$

and here C_0 introduced by ordinary integration is an absolute constant, while C_1 introduced by the performance of the operation Σ is a periodical constant.

A superfluity among the arbitrary constants, but a superfluity which does not affect their arbitrariness, is always to be *presumed* when the inverse operations by which they are introduced are at a subsequent stage of the process of solution followed by the corresponding *direct* operations. The particular observations of Chap. XVII. Art. 4. (*Differential Equations*) on this subject admit of a wider application.

Secondly, let n be 0 or a negative integer.

It is here desirable to change the sign of n so as to express the given equation in the form

$$\frac{du_1}{dx} - a\frac{du}{dx} + (x - n)\, u_1 - axu = 0,$$

while its symbolical solution (A) becomes

$$u = (D - a)^{-n-1} \epsilon^{\frac{-x^2}{2}} \left(\frac{d}{dx}\right)^{-1} \epsilon^{\frac{x^2}{2}} (D - a)^n\, 0.$$

And in both n is 0 or a positive integer.

Now since $(D - a)^n\, 0 = 0$, and $\left(\dfrac{d}{dx}\right)^{-1} 0 = C$, we have

$$u = (D - a)^{-n-1}\, C\epsilon^{\frac{-x^2}{2}}$$

$$= C\, (D - a)^{-n-1}\, \epsilon^{\frac{-x^2}{2}} + (D - a)^{-n-1}\, 0$$

$$= Ca^{x-n-1}\, \Sigma^{n+1}\, a^{-x}\, \epsilon^{\frac{-x^2}{2}} + a^{x-n-1}\, \Sigma^{n+1} 0$$

$$= C_1 a^x\, \Sigma^{n+1} a^{-x}\, \epsilon^{\frac{-x^2}{2}} + a^x\, (c_0 + c_1 x \ldots + c_n x^n).$$

But here, while the absolute constant C_1 is arbitrary, the $n + 1$ periodical constants $c_0, c_1 \ldots c_n$ are connected by n relations which must be determined by substitution of the above unreduced value of u in the given equation.

The general expression of these relations is somewhat complex; but in any particular case they may be determined without difficulty.

Thus if $a = 1$, $n = 1$, it will be found that

$$u = C_1 \Sigma^2 a^{-x} \epsilon^{-\frac{x^2}{2}} + C_2 (1 - x).$$

If $a = 1$, $n = 2$, we shall have

$$u = C_1 \Sigma^3 a^{-x} \epsilon^{-\frac{x^2}{2}} + C_2 \left(1 - x + \frac{x^2}{3}\right),$$

and so on.

The two general solutions may be verified, though not easily, by substitution in the original equation.

12. The same principles of solution are applicable to equations of mixed partial differences as to equations of partial differences. If Δ_x and $\frac{d}{dy}$ are the symbols of pure operation involved, and if, replacing one of these by a constant m, the equation becomes either a pure differential equation or a pure equation of differences with respect to the other, then it is only necessary to replace in the solution of that equation m by the symbol for which it stands, to effect the corresponding change in the arbitrary constant, and then to interpret the result.

Ex. 11. $\Delta_x u - a \dfrac{du}{dy} = 0.$

Replacing $\frac{d}{dy}$ by m, and integrating, we have

$$u = c (1 + am)^x.$$

Hence the symbolic solution of the given equation is

$$u = \left(1 + a \frac{d}{dy}\right)^x \phi(y)$$

$$= a^x \left(\frac{d}{dy} + \frac{1}{a}\right)^x \phi(y)$$

$$= a^x \epsilon^{-\frac{y}{a}} \left(\frac{d}{dy}\right)^x \epsilon^{\frac{y}{a}} \phi(y),$$

$$= a^x \epsilon^{-\frac{y}{a}} \left(\frac{d}{dy}\right)^x \psi(y),$$

$\psi(y)$ being an arbitrary function of y.

Ex. 12.   Given $u_{x+1, y} - \dfrac{d}{dy} u_{x, y} = V_{x, y}$.

Treating $\dfrac{d}{dy}$ as a constant, the symbolic solution is

$$u_{x, y} = \left(\frac{d}{dy}\right)^{x-1} \Sigma \left(\frac{d}{dy}\right)^{-x} V_{x, y} + \left(\frac{d}{dy}\right)^x \phi(y),$$

Σ having reference to x.   No constants need to be introduced in performing the integrations implied by $\left(\dfrac{d}{dy}\right)^{-x}$.

Ex. 13.   Given $u_{x+2} - 3x \dfrac{du_{x+1}}{dy} + 2x\,(x-1)\dfrac{d^2 u_x}{dy^2} = 0$.

Let $u_x = 1 . 2 \ldots (x-2)\, v_x$, then

$$v_{x+2} - 3\,\frac{dv_{x+1}}{dy} + 2\,\frac{d^2 v_x}{dy^2} = 0,$$

or

$$\left\{D^2 - 3D_x \frac{d}{dy} + 2\left(\frac{d}{dy}\right)^2\right\} v_x = 0,$$

or

$$\left(D_x - \frac{d}{dy}\right)\left(D_x - 2\,\frac{d}{dy}\right) v_x = 0,$$

whence by resolution and integration

$$v_x = \left(\frac{d}{dy}\right)^x \phi(y) + \left(2\,\frac{d}{dy}\right)^x \psi(y),$$

$$u_x = 1 . 2 \ldots (x-2)\left\{\left(\frac{d}{dy}\right)^x \phi(y) + 2^x \left(\frac{d}{dy}\right)^x \psi(y)\right\}.$$

Ex. 14. $u_{x+2} - 3\dfrac{du_{x+1}}{dy} + 2\dfrac{d^2 u_x}{dy^2} = V$, where V is a function of x and y.

Here we have

$$\left\{ D_x^2 - 3D_x \frac{d}{dy} + 2 \left(\frac{d}{dy}\right)^2 \right\} u_x = V;$$

$$\therefore\ u = \left\{ \left(D_x - \frac{d}{dy}\right) \left(D_x - 2\frac{d}{dy}\right)^{-1} \right\} V$$

$$= \frac{d}{dy} \left(D_x - 2\frac{d}{dy}\right)^{-1} V - \frac{d}{dy} \left(D_x - \frac{d}{dy}\right)^{-1} V$$

$$= \frac{d}{dy} \left(2\frac{d}{dy}\right)^{x-1} \Sigma \left(2\frac{d}{dy}\right)^{-x} V - \frac{d}{dy} \left(\frac{d}{dy}\right)^{x-1} \Sigma \left(\frac{d}{dy}\right)^{-x} V.$$

The complementary part of the value of u introduced by the performance of Σ will evidently be

$$2^x \left(\frac{d}{dy}\right)^x \phi(y) + \left(\frac{d}{dy}\right)^x \psi(y).$$

But in particular cases the difficulties attending the reduction of the general solution may be avoided.

Thus, representing V by V_x, we have, as a particular solution

$$u_x = \left\{ D_x^2 - 3D_x \frac{d}{dy} + 2 \left(\frac{d}{dy}\right)^2 \right\}^{-1} V_x$$

$$= \left(D_x^{-2} + 3D_x^{-3} \frac{d}{dy} + 7D_x^{-4} \frac{d^2}{dy^2} + \&c. \right) V_x$$

$$= V_{x-2} + 3 \frac{dV_{x-3}}{dy} + 7 \frac{d^2 V_{x-4}}{dy^2} + \&c.,$$

which terminates if V_x is rational and integral with respect to y. The complement must then be added.

Thus the complete solution of the given equation when

$$V = F(x) + y,$$

is $u = F(x-2) + y + 3 + 2^x \left(\frac{d}{dy}\right)^x \phi(y) + \left(\frac{d}{dy}\right)^x \psi(y).$

Simultaneous Equations.

13.   Whether the equation be one of ordinary differences, or of mixed differences, the principle of the method of solution is the same.   We must, by the performance of the operations Δ_x and $\dfrac{d}{dx}$, obtain a system of derived equations sufficient to enable us by elimination to deduce a final equation involving only one of the dependent variables with its differences or differentials.   The integrations of this will give the general value of that variable, and the equations employed in the process of elimination will enable us to express each other dependent variable by means of it.   If the coefficients are constant we may simply separate the symbols and effect the eliminations as if those symbols were algebraic.

Ex. 15.   Given $u_{x+1} - a^2 x v_x = 0$, $v_{x+1} - x u_x = 0$.

From the first we have

$$u_{x+2} - a^2 (x + 1) v_{x+1} = 0.$$

Hence, eliminating v_{x+1} by the second,

$$u_{x+2} - a^2 x (x + 1) u_x = 0,$$

the solution of which is

$$u_x = 1 . 2 \ldots (x-1) \{ C_1 a^x + C_2 (-a)^x \}.$$

Then by the first equation

$$v_x = \frac{u_{x+1}}{a^2 x} = \frac{1 . 2 \ldots (x-1)}{a^2} \{ C_1 a^{x+1} + C_2 (-a)^{x+1} \}.$$

Ex. 16.   Given $u_{x+2} + 2 v_{x+1} - 8 u_x = a^x$,

$$v_{x+2} - u_{x+1} - 2 v_x = a^{-x}.$$

Separating the symbols

$$(D^2 - 8) u_x + 2 D v_x = a^x,$$
$$D u_x - (D^2 - 2) v_x = - a^{-x}.$$

Eliminating v_x and reducing

$$(D^2 - 4)^2 u_x = (a^2 - 2)a^x - 2a^{-x-1}.$$

The complete integral of which is

$$u_x = \frac{a^2 - 2}{(a^2 - 4)^2} a^x - \frac{2}{a} \frac{a^{-x}}{(a^{-2} - 4)^2}$$

$$+ (c_0 + c_1 x) 2^x + (c_2 + c_3 x) (-2)^x.$$

Moreover by the first of the given equations

$$v_x = \frac{a^{x-1} + 8u_{x-1} - u_{x+1}}{2}.$$

EXERCISES.

1. Integrate the simultaneous equations

$$u_{x+1} - v_x = 2m (x + 1) \dots\dots\dots\dots(1),$$
$$v_{x+1} - u_x = - 2m (x + 1) \dots\dots\dots\dots(2).$$

2. Integrate the system $u_{x+1} + (- 1)^x v_x = 0,$
$$v_{x+1} + (- 1)^x u_x = 0.$$

3. Integrate the system $v_{x+1} - u_x = (l - m) x,$
$$w_{x+1} - v_x = (m - n) x,$$
$$u_{x+1} - w_x = (n - l) x.$$

4. Integrate $\Delta_x u_{x,y} - a \dfrac{d}{dy} u_{x,y} = 0.$

5. $u_{x+2,y} - a \dfrac{d}{dy} u_{x+1,y} + b \dfrac{d^2}{dy^2} u_{x,y} = 0.$

6. $u_{x+1,y} - u_{x,y+1} = x + y.$

7. $u_{x+1,y+n} - u_{x,y} = a^{x-y}.$

8. Determine $u_{x,t}$ from the equation

$$c^2 \frac{d^2}{dt^2} u_{x+2,t} = \Delta^2 u_{x,t},$$

where Δ affects x only; and, assuming as initial conditions

$$u_{x,0} = ax + b, \quad \frac{d}{d0} u_{x,0} = a' r^x,$$

shew that

$$\frac{d}{dt} u_{x,t} = A \lambda^x (\mu^t + \mu^{-t}),$$

where A, λ and μ are constants (*Cambridge Problems*).

CHAPTER XI.

OF THE CALCULUS OF FUNCTIONS.

1. THE Calculus of Functions in its purest form is distinguished by this, viz. that it recognises no other operations than those which are termed functional. In the state to which it has been brought, more especially by the labors of Mr Babbage, it is much too extensive a branch of analysis to permit of our attempting here to give more than a general view of its objects and its methods. But it is proper that it should be noticed, 1st, because the Calculus of Finite Differences is but a particular form of the Calculus of Functions; 2ndly, because the methods of the more general Calculus are in part an application, in part an extension of those of the particular one.

In the notation of the Calculus of Functions, $\phi \{\psi (x)\}$ is usually expressed in the form $\phi\psi x$, brackets being omitted except when their use is indispensable. The expressions $\phi\phi x$, $\phi\phi\phi x$ are, by the adoption of indices, abbreviated into $\phi^2 x$, $\phi^3 x$, &c. As a consequence of this notation we have $\phi^0 x = x$ independently of the form of ϕ. The inverse form ϕ^{-1} is, it must be remembered, *defined* by the equation

$$\phi\phi^{-1} x = x \quad\dots\dots\dots\dots\dots\dots (1).$$

Hence ϕ^{-1} may have different forms corresponding to the same form of ϕ. Thus if

$$\phi x = x^2 + ax,$$

we have, putting $\phi x = t$,

$$x = \phi^{-1} t = - \frac{a \pm \sqrt{(a^2 + 4t)}}{2},$$

and ϕ^{-1} has two forms.

The problems of the Calculus of Functions are of two kinds, viz.

1st. Those in which it is required to determine a functional form equivalent to some known combination of known forms; e. g. from the form of ψx to determine that of $\psi^n x$.

2ndly. Those which involve the solution of functional equations, i. e. the determination of an unknown function from the conditions to which it is subject, not as in the previous case from the known mode of its composition.

We may properly distinguish these problems as direct and inverse. Problems will of course present themselves in which the two characters meet.

Direct Problems.

2.   Given the form of ψx, required that of $\psi^n x$.

There are cases in which this problem can be solved by successive substitution.

Ex. 1.   Thus, if $\psi x = x^a$, we have
$$\psi \psi x = (x^a)^a = x^{a^2},$$
and generally
$$\psi^n x = x^{a^n}.$$

Again, if on determining $\psi^2 x$, $\psi^3 x$ as far as convenient it should appear that some one of these assumes the particular form x, all succeeding forms will be determined.

Ex. 2.   Thus if $\psi x = 1 - x$, we have
$$\psi^2 x = 1 - (1 - x) = x.$$

Hence $\psi^n x = 1 - x$ or x according as n is odd or even.

Ex. 3.   If $\psi x = \dfrac{1}{1 - x}$, we find
$$\psi^2 x = \frac{x - 1}{x}, \; \psi^3 x = x.$$

Hence $\psi^n x = x, \dfrac{1}{1-x}$ or $\dfrac{x-1}{x}$ according as on dividing n by 3 the remainder is 0, 1 or 2.

Functions of the above class are called periodic, and are distinguished in order according to the number of distinct forms to which $\psi^n x$ gives rise for integer values of n. The function in Ex. 2 is of the second, that in Ex. 3 of the third, order.

Theoretically the solution of the general problem may be made to depend upon that of an equation of differences of the first order. For assume

$$\psi^n x = t_n, \ \psi^{n+1} x = t_{n+1} \ \dots\dots\dots (2).$$

Then, since $\psi^{n+1} x = \psi \psi^n x$, we have

$$t_{n+1} = \psi(t_n) \ \dots\dots\dots\dots (3).$$

The arbitrary constant in the solution of this equation may be determined by the condition $t_1 = \psi x$, or by the still prior condition

$$t_0 = \psi^0 x = x \ \dots\dots\dots\dots (4).$$

It will be more in analogy with the notation of the other chapters of this work if we present the problem in the form: Given ψt, required $\psi^x t$, thus making x the independent variable of the equation of differences.

Ex. 4.   Given $\psi t = a + bt$, required $\psi^x t$.

Assuming $\psi^x t = u_x$ we have

$$u_{x+1} = a + bu_x,$$

the solution of which is

$$u_x = cb^x + \frac{a}{1-b}.$$

Now $u_0 = \psi^0 t = t$, therefore

$$t = c + \frac{a}{1-b}.$$

Hence determining c we find on substitution

$$u_x = a\,\frac{b^x - 1}{b - 1} + b^x t \dots\dots\dots\dots (5),$$

the expression for $\psi^x t$ required.

Ex. 5. Given $\psi t = \dfrac{a}{b + t}$, required $\psi^x t$.

Assuming $\psi^x t = u_x$ we have

$$u_{x+1} = \frac{a}{b + u_x},$$

or $\qquad\qquad u_x u_{x+1} + b u_{x+1} = a.$

This belongs to the third of the forms treated in Chap. VII. Art. 9. Assume therefore

$$u_x + b = \frac{v_{x+1}}{v_x},$$

then $\qquad\qquad v_{x+2} - b v_{x+1} - a v_x = 0,$
the solution of which is

$$v_x = c_1 \alpha^x + c_2 \beta^x,$$

α and β being the roots of the equation

$$m^2 - bm - a = 0.$$

Hence $\qquad\qquad u_x = \dfrac{c_1 \alpha^{x+1} + c_2 \beta^{x+1}}{c_1 \alpha^x + c_2 \beta^x} - b;$

or, putting C for $\dfrac{c_2}{c_1}$ and $\alpha + \beta$ for b, and reducing,

$$u_x = -\alpha\beta\,\frac{\alpha^{x-1} + C\beta^{x-1}}{\alpha^x + C\beta^x} \dots\dots\dots\dots (6).$$

Now $u_0 = \psi^0 t = t$, therefore

$$t = -\alpha\beta\,\frac{\alpha^{-1} + C\beta^{-1}}{1 + C}$$

$$= -\frac{\beta + C\alpha}{1 + C},$$

whence
$$C = -\frac{t+\beta}{t+\alpha};$$

and, substituting in (6),

$$u_x = -\alpha\beta \frac{\alpha^x - \beta^x + (\alpha^{x-1} - \beta^{x-1})\, t}{\alpha^{x+1} - \beta^{x+1} + (\alpha^x - \beta^x)\, t} \quad \dots\dots\dots (7),$$

the expression for $\psi^x t$ required.

Since in the above example $\psi t = \dfrac{a}{b+t}$, we have, by direct substitution,

$$\psi^2 t = \frac{a}{b + \psi t} = \frac{a}{b + \dfrac{a}{b+t}},$$

and continuing the process and expressing the result in the usual notation of continued fractions,

$$\psi^x t = \frac{a}{b+} \frac{a}{b+} \frac{a}{b+} \dots \frac{a}{b+t},$$

the number of simple fractions being x. Of the value of this continued fraction the right-hand member of (7) is therefore the finite expression. And the method employed shews how the calculus of finite differences may be applied to the finite evaluation of various other functions involving definite repetitions of given functional operations.

Ex. 6.   Given $\psi t = \dfrac{a+bt}{c+et}$, required $\psi^x t$.

Assuming as before $\psi^x t = u_x$, we obtain as the equation of differences

$$e u_x u_{x+1} + c u_{x+1} - b u_x - a = 0 \quad \dots\dots\dots\dots (8),$$

and applying to this the same method as before, we find

$$u_x = \frac{\alpha^{x+1} + C\beta^{x+1}}{\alpha^x + C\beta^x} - \frac{c}{e} \quad \dots\dots\dots\dots\dots (9),$$

α and β being the roots of

$$e^2 m^2 - (b+c)\, em + bc - ae = 0 \quad \dots\dots\dots\dots (10);$$

and in order to satisfy the condition $u_0 = t$,

$$C = -\frac{e(t-\alpha)+c}{e(t-\beta)+c} \dots\dots\dots\dots\dots (11).$$

When α and β are imaginary, the exponential forms must be replaced by trigonometrical ones. We may, however, so integrate the equation (8) as to arrive directly at the trigonometrical solution.

For let that equation be placed in the form

$$\left(u_x + \frac{c}{e}\right)\left(u_{x+1} - \frac{b}{e}\right) + \frac{bc-ae}{e^2} = 0.$$

Then assuming $u_x = t_x + \dfrac{b-c}{2e}$, we have

$$\left(t_x + \frac{b+c}{2e}\right)\left(t_{x+1} - \frac{b+c}{2e}\right) + \frac{bc-ae}{e^2} = 0,$$

or $\qquad t_x t_{x+1} + \mu(t_{x+1} - t_x) + \nu^2 = 0 \dots\dots\dots (12),$

in which $\qquad \mu = \dfrac{b+c}{2e}, \quad \nu^2 = \dfrac{bc-ae}{e^2} - \dfrac{(b+c)^2}{4e^2} \dots\dots (13).$

Hence $\qquad \dfrac{t_{x+1}-t_x}{\nu^2 + t_x t_{x+1}} = -\dfrac{1}{\mu},$

or, assuming $t_x = \nu s_x$,

$$\frac{s_{x+1}-s_x}{1 + s_x s_{x+1}} = \frac{-\nu}{\mu},$$

the integral of which is

$$s_x = \tan\left(C - x\tan^{-1}\frac{\nu}{\mu}\right).$$

But $t_x = \nu s_x$ and $u_x = t_x + \mu'$, where

$$\mu' = \frac{b-c}{2e} \dots\dots\dots\dots\dots (14).$$

Hence $\qquad u_x = \nu\tan\left(C - x\tan^{-1}\dfrac{\nu}{\mu}\right) + \mu' \dots\dots\dots (15),$

the general integral.

Now the condition $u_0 = t$ gives

$$t = \nu \tan C + \mu'.$$

Hence determining C we have, finally,

$$\psi^x t = \nu \tan \left(\tan^{-1} \frac{t - \mu'}{\nu} - x \tan^{-1} \frac{\nu}{\mu} \right) + \mu' \ldots\ldots\ldots(16),$$

for the general expression of $\psi^x t$.

This expression is evidently reducible to the form

$$\frac{A + Bt}{C + Et},$$

the coefficients A, B, C, E being functions of x.

Reverting to the exponential form of $\psi^x t$ given in (9), it appears from (10) that it is real if the function

$$\frac{(b + c)^2}{e^2} - 4 \frac{bc - ae}{e^2}$$

is positive. But this is the same as $-4\nu^2$. The trigonometrical solution therefore applies when the expression represented by ν^2 is positive, the exponential one when it is negative.

In the case of $\nu = 0$ the equation of differences (12) becomes

$$t_x t_{x+1} + \mu (t_{x+1} - t_x) = 0,$$

or

$$\frac{1}{t_{x+1}} - \frac{1}{t_x} = \frac{1}{\mu},$$

the integral of which is

$$t_x = \frac{x}{\mu} + C.$$

Determining the constant as before we ultimately get

$$\psi^x t = \frac{\mu'^2 x - (\mu + \mu' x) t}{\mu' x - \mu - xt} \ldots\ldots\ldots\ldots(17),$$

a result which may also be deduced from the trigonometrical solution by the method proper to indeterminate functions.

Periodical Functions.

3. It is thus seen, and it is indeed evident *a priori*, that in the above cases the form of $\psi^x t$ is similar to that of ψt, but with altered constants. The only functions which are known to possess this property are

$$\frac{a+bt}{c+et} \text{ and } at^c.$$

On this account they are of great importance in connexion with the general problem of the determination of the possible forms of periodical functions, particular examples of which will now be given.

Ex. 7. Under what conditions is $a + bt$ a periodical function of the xth order?

By Ex. 4 we have

$$\psi^x t = a\,\frac{b^x-1}{b-1} + b^x t,$$

and this, for the particular value of x in question, must reduce to t. Hence

$$a\,\frac{b^x-1}{b-1} = 0, \quad b^x = 1,$$

equations which require that b should be any x^{th} root of unity except 1 when a is not equal to 0, and any x^{th} root of unity when a is equal to 0.

Hence if we confine ourselves to real forms the only periodic forms of $a + bt$ are t and $a - t$, the former being of every order, the latter of every even order.

Ex. 8. Required the conditions under which $\dfrac{a+bt}{c+et}$ is a periodical function of the xth order.

In the following investigation we exclude the supposition of $e = 0$, which merely leads to the case last considered.

Making then in (16) $\psi^x t = t$, we have

$$t = \mu' + \nu \tan \left(\tan^{-1} \frac{t - \mu'}{\nu} - x \tan^{-1} \frac{\nu}{\mu} \right) \dots\dots\dots (18),$$

or $$\frac{t - \mu'}{\nu} = \tan \left(\tan^{-1} \frac{t - \mu'}{\nu} - x \tan^{-1} \frac{\nu}{\mu} \right),$$

an equation which, with the exception of a particular case to be noted presently, is satisfied by the assumption

$$x \tan^{-1} \frac{\nu}{\mu} = i\pi,$$

i being an integer.   Hence we have

$$\frac{\nu}{\mu} = \tan \frac{i\pi}{x} \dots\dots\dots\dots\dots\dots (19),$$

or, substituting for ν and μ their values from (13),

$$\frac{4 (bc - ae)}{(b + c)^2} - 1 = \tan^2 \frac{i\pi}{x},$$

whence we find

$$e = - \frac{b^2 - 2bc \cos \dfrac{2i\pi}{x} + c^2}{4a \cos^2 \dfrac{i\pi}{x}} \dots\dots\dots\dots\dots (20).$$

The case of exception above referred to is that in which $\nu = 0$, and in which therefore, as is seen from (19), i is a multiple of x. For the assumption $\nu = 0$ makes the expression for t given in (18) indeterminate, the last term assuming the form $0 \times \infty$. If the true limiting value of that term be found in the usual way, we shall find for t the same expression as was obtained in (17) by direct integration. But that expression would lead merely to $x = 0$ as the condition of periodicity, a condition which however is satisfied by all functions whatever, in virtue of the equation $\phi^0 t = t$.

The solution (9) expressed in exponential forms does not lead to any condition of periodicity when a, b, c, e are real quantities.

We conclude that the conditions under which $\dfrac{a + bt}{c + et}$, *when not of the form $A + Bt$, is a periodical function of the xth order, are expressed by* (20), i *being any integer which is not a multiple of x**.

4. From any given periodical function an infinite number of others may be deduced by means of the following theorem.

THEOREM.   If ft be a periodical function, then $\phi f \phi^{-1} t$ is also a periodical function of the same order.

For let $\qquad \phi f \phi^{-1} t = \psi t,$

then $\qquad\qquad \psi^2 t = \phi f \phi^{-1} \phi f \phi^{-1} t$
$$= \phi f^2 \phi^{-1} t.$$

And continuing the process of substitution
$$\psi^n t = \phi f^n \phi^{-1} t.$$

Now, if ft be periodic of the nth order, $f^n t = t$, and
$$f^n \phi^{-1} t = \phi^{-1} t.$$

Hence $\qquad\qquad \psi^n t = \phi \phi^{-1} t = t.$

Therefore ψt is periodic of the nth order.

Thus, it being given that $1 - t$ is a periodic function of t of the second order, other such functions are required.

Represent $1 - t$ by ft.

Then if $\phi t = t^2$,
$$\phi f \phi^{-1} t = (1 - \sqrt{t})^2.$$

If $\phi t = \sqrt{t}$,
$$\phi f \phi^{-1} t = (1 - t^2)^{\frac{1}{2}}.$$

These are periodic functions of the second order; and the number might be indefinitely multiplied.

The system of functions included in the general form $\phi f \phi^{-1} t$ have been called the *derivatives* of the function ft.

---

* I am not aware that the limitation upon the integral values of i has been noticed before.

Functional Equations.

5. The most general definition of a functional equation is that it expresses a relation arising from the *forms* of functions; a relation therefore which is independent of the particular values of the subject variable. The object of the solution of a functional equation is the discovery of an unknown form from its relation thus expressed with forms which are known.

The nature of functional equations is best seen from an example of the mode of their genesis.

Let $f(x, c)$ be a given function of x and c, which considered as a function of x, may be represented by ϕx, then

$$\phi x = f(x, c),$$

and changing x into any *given* function ψx,

$$\phi \psi x = f(\psi x, c).$$

Eliminating c between these two equations we have a result of the form

$$F(x, \phi x, \phi \psi x) = 0 \dots\dots\dots\dots(1).$$

This is a functional equation, the object of the solution of which would be the discovery of the form ϕ, those of F and ψ being given.

It is evident that neither the above process nor its result would be affected if c instead of being a constant were a function of x which did not change its form when x was changed into ψx. Thus if we assume as a primitive equation

$$\phi(x) = cx + \frac{1}{c} \dots\dots\dots\dots\dots\dots(a),$$

and change x into $-x$, we have

$$\phi(-x) = -cx + \frac{1}{c}.$$

Eliminating c we have, on reduction,

$$\{\phi(x)\}^2 - \{\phi(-x)\}^2 = 4x,$$

a functional equation of which (a) constitutes the complete primitive. In that primitive we may however interpret c as an arbitrary *even* function of x, the only condition to which it is subject being that it shall not change on changing x into $-x$. Thus we should have as *particular* solutions

$$\phi(x) = x \cos x + \frac{1}{\cos x},$$

$$\phi(x) = x^3 + \frac{1}{x^2},$$

these being obtained by assuming $c = \cos x$ and x^2 respectively.

Equations of differences are a particular species of functional equations, the elementary functional change being that of x into $x+1$. And the most general method of solving functional equations of all species, consists in reducing them to equations of differences. Laplace has given such a method, which we shall exemplify upon the equation

$$F(x, \phi\psi x, \phi\chi x) = 0 \ldots\ldots\ldots\ldots\ldots(2),$$

the forms of ψ and χ being known and that of ϕ sought. But though we shall consider the above equation under its general form, we may remark that it is reducible to the simpler form (1). For, the form of ψ being known, that of ψ^{-1} may be presumed to be known also. Hence if we put $\psi x = z$ and $\chi\psi^{-1}z = \psi_1 z$, we have

$$F(\psi^{-1}z, \phi z, \phi\psi_1 z) = 0,$$

and this, since ψ^{-1} and ψ_1 are known, is reducible to the general form (1).

Now resuming (2) let

$$\left.\begin{array}{ll} \psi x = u_t, & \chi x = u_{t+1} \\ \phi\psi x = v_t, & \phi\chi x = v_{t+1} \end{array}\right\} \ldots\ldots\ldots\ldots (3).$$

Hence v_t and u_t being connected by the relation

$$v_t = \phi u_t \ldots\ldots\ldots\ldots\ldots\ldots (4),$$

the form of ϕ will be determined if we can express v_t as a function of u_t.

Now the first two equations of the system give on eliminating x an equation of differences of the form

$$u_{t+1} = fu_t \dots\dots\dots\dots\dots (5),$$

the solution of which will determine u_t, therefore ψx, therefore, by inversion, x as a function of t. This result, together
with the last two equations of the system (3) will convert the
given equation (2) into an equation of differences of the
first order between t and v_t, the solution of which will determine v_t as a function of t, therefore as a function of u_t since
the form of u_t has already been determined. But this determination of v_t as a function of u_t is equivalent, as has been seen,
to the determination of the form of ϕ.

Ex. 9.   Let the given equation be $\phi(mx) - a\phi(x) = 0$.

Then assuming

$$\left.\begin{array}{ll} x = u_t, & mx = u_{t+1} \\ \phi(x) = v_t, & \phi(mx) = v_{t+1} \end{array}\right\} \dots\dots\dots (a),$$

we have from the first two

$$u_{t+1} - mu_t = 0,$$

the solution of which is

$$u_t = Cm^t \dots\dots\dots\dots (b).$$

Again, by the last two equations of (a) the given equation
becomes

$$v_{t+1} - av_t = 0,$$

whence

$$v_t = C'a^t \dots\dots\dots\dots (c).$$

Eliminating t between (b) and (c), we have

$$v_t = C'a^{\frac{\log u_t - \log C}{\log m}}.$$

Hence replacing u_t by x, v_t by ϕx, and $C'a^{-\frac{\log C}{\log m}}$ by C_1, we
have

$$\phi x = C_1 a^{\frac{\log x}{\log m}} \dots\dots\dots\dots (d).$$

And here C_1 must be interpreted as any function of x which does not change on changing x into mx.

If we attend strictly to the analytical origin of C_1 in the above solution we should obtain for it the expression

$$a_0 + a_1 \cos\left(2\pi \frac{\log x}{\log m}\right) + a_2 \cos\left(4\pi \frac{\log x}{\log m}\right) + \&c.$$

$$+ b_1 \sin\left(2\pi \frac{\log x}{\log m}\right) + b_2 \sin\left(4\pi \frac{\log x}{\log m}\right) + \&c.$$

a_0, a_1, b_1, &c. being *absolute* constants. But it suffices to adopt the simpler definition given above, and such a course we shall follow in the remaining examples.

Ex. 10.   Given $\phi\left(\dfrac{1+x}{1-x}\right) - a\phi(x) = 0.$

Assuming

$$x = u_t, \quad \frac{1+x}{1-x} = u_{t+1},$$

$$\phi(x) = v_t, \quad \phi\left(\frac{1+x}{1-x}\right) = v_{t+1},$$

we have

$$u_{t+1} = \frac{1+u_t}{1-u_t},$$

or

$$u_t u_{t+1} - u_{t+1} + u_t + 1 = 0.$$

The solution of which is

$$u_t = \tan\left(C + \frac{\pi}{4}t\right).$$

Again we have

$$v_{t+1} - av_t = 0,$$

whence

$$v_t = C'a^t.$$

Hence replacing u_t by x, v_t by $\phi(x)$, and eliminating t,

$$\phi(x) = C_1 a^{\frac{4}{\pi}\tan^{-1}x},$$

C_1 being any function of x which does not change on changing x into $\dfrac{1+x}{1-x}$.

6. Linear functional equations of the form

$$\phi\psi^n x + a_1\phi\psi^{n-1}x + a_2\phi\psi^{n-2}x \ldots + a_n\phi\,(x) = X \ \ldots\ldots (6),$$

where $\psi\,(x)$ is a known function of x, may be reduced to the preceding form.

For let π be a symbol which operating on any function $\phi\,(x)$ has the effect of converting it into $\phi\psi\,(x)$. Then the above equation becomes

$$\pi^n\phi\,(x) + a_1\pi^{n-1}\phi\,(x) \ldots + a_n\phi\,(x) = X,$$

or

$$(\pi^n + a_1\pi^{n-1} \ldots + a_n)\,\phi\,(x) = X \ \ldots\ldots\ldots (7).$$

It is obvious that π possesses the distributive property expressed by the equation

$$\pi\,(u+v) = \pi u + \pi v,$$

and that it is commutative with constants so that

$$\pi a u = a\pi u.$$

Hence we are permitted to reduce (7) in the following manner, viz.

$$\phi\,(x) = (\pi^n + a_1\pi^{n-1} \ldots + a_n)^{-1} X$$
$$= \{N_1\,(\pi - m_1)^{-1} + N_2\,(\pi - m_2)^{-1} \ldots\}\,X \ \ldots\ldots (8),$$

$m_1,\ m_2 \ldots$ being the roots of

$$m^n + a_1 m^{n-1} \ldots + a_n = 0 \ \ldots\ldots\ldots\ldots (9),$$

and $N_1, N_2 \ldots$ having the same values as in the analogous resolution of rational fractions.

Now if $(\pi - m)^{-1} X = \phi\,(x)$, we have

$$(\pi - m)\,\phi\,(x) = X,$$

or $\qquad \phi\psi(x) - m\phi(x) = X,$

to which Laplace's method may be applied.

Ex. 11.   Given $\phi(m^2x) + a\phi(mx) + b\phi(x) = x^n.$

Representing by α and β the roots of $x^2 + ax + b = 0$, the solution is

$$\phi(x) = \frac{x^n}{m^{2n} + am^n + b} + Cx^{\frac{\log\alpha}{\log m}} + C'x^{\frac{\log\beta}{\log m}},$$

C and C' being functions of x unaffected by the change of x into mx.

Here we may notice that just as in linear differential equations and in linear equations of differences, and for the same reason, viz. the distributive character of the symbol π, the complete value of $\phi(x)$ consists of two portions, viz. of any particular value of $\phi(x)$ together with what would be its complete value were $X = 0$. This is seen in the above example.

7.   There are some cases in which particular solutions of functional equations, more especially if the known functions involved in the equations are periodical, may be obtained with great ease. The principle of their solution is as follows.

Supposing the given equation to be

$$F(x, \phi x, \phi\psi x) = 0 \quad \dots\dots\dots\dots\dots (10),$$

and let ψx be a periodical function of the second order. Then changing x into ψx, and observing that $\psi^2 x = x$, we have

$$F(\psi x, \phi\psi x, \phi x) = 0 \quad \dots\dots\dots\dots (11).$$

Eliminating $\phi\psi x$ the resulting equation will determine ϕx as a function of x and ψx, and therefore since ψx is supposed known, as a function of x.

If ψx is a periodical function of the third order, it would be necessary to effect the substitution twice in succession, and then to eliminate $\phi\psi x$, and $\phi\psi^2 x$; and so on according to the order of periodicity of ψx.

Ex. 12.   Given $(\phi x)^2 \phi \dfrac{1-x}{1+x} = a^2 x.$

The function $\dfrac{1-x}{1+x}$ is periodic of the second order. Change

then x into $\dfrac{1-x}{1+x}$, and we have

$$\left(\phi \frac{1-x}{1+x}\right)^2 \phi x = a^2 \frac{1-x}{1+x}.$$

Hence, eliminating $\phi \dfrac{1-x}{1+x}$, we find

$$\phi x = a^{\frac{2}{3}} x^{\frac{2}{3}} \left(\frac{1+x}{1-x}\right)^{\frac{1}{3}}$$

as a particular solution. (Babbage, *Examples of Functional Equations*, p. 7).

This method fails if the process of substitution does not yield a number of *independent* equations sufficient to enable us to effect the elimination. Thus, supposing ψx a periodical function of the second order, it fails for equations of the form

$$F(\phi x, \phi \psi x) = 0,$$

if symmetrical with respect to ϕx and $\phi \psi x$. In such cases we must either, with Mr Babbage, treat the given equation as a particular case of some more general equation which is unsymmetrical, or we must endeavour to solve it by some more general method like that of Laplace.

Ex. 13.   Given

$$(\phi x)^2 + \left\{\phi \left(\frac{\pi}{2} - x\right)\right\}^2 = 1.$$

This is a particular case of the more general equation

$$(\phi x)^2 + m \left\{\phi \left(\frac{\pi}{2} - x\right)\right\}^2 = 1 + n \chi x,$$

m and n being constants which must be made equal to 1 and 0 respectively, and χx being an arbitrary function of x.

Changing x into $\frac{\pi}{2} - x$, we have

$$\left\{\phi\left(\frac{\pi}{2} - x\right)\right\}^2 + m\left\{\phi(x)\right\}^2 = 1 + n\chi\left(\frac{\pi}{2} - x\right).$$

Eliminating $\phi\left(\frac{\pi}{2} - x\right)$ from the above equations, we find

$$(1 - m^2)\left\{\phi(x)\right\}^2 = 1 - m + n\left\{\chi x - m\chi\left(\frac{\pi}{2} - x\right)\right\}.$$

Therefore

$$\left\{\phi(x)\right\}^2 = \frac{1}{1 + m} + \frac{n}{1 - m^2}\left\{\chi x - m\chi\left(\frac{\pi}{2} - x\right)\right\}.$$

Now if m become 1 and n become 0, independently, the fraction $\frac{n}{1 - m^2}$ becomes indeterminate, and may be replaced by an arbitrary constant c. Thus we have

$$\left\{\phi(x)\right\}^2 = \frac{1}{2} + c\chi(x) - c\chi\left(\frac{\pi}{2} - x\right);$$

whence, merging c in the arbitrary function,

$$\phi(x) = \left\{\frac{1}{2} + \chi(x) - \chi\left(\frac{\pi}{2} - x\right)\right\}^{\frac{1}{2}} \quad\ldots\ldots\ldots\ (12).$$

The above is in effect Mr Babbage's solution, excepting that, making m and n dependent, he finds a particular value for the fraction which in the above solution becomes an arbitrary constant.

Let us now solve the equation by Laplace's method. Let $\left\{\phi(x)\right\}^2 = \psi x$, and we have

$$\psi(x) + \psi\left(\frac{\pi}{2} - x\right) = 1.$$

Hence assuming

$$x = u_t, \quad \frac{\pi}{2} - x = u_{t+1},$$

$$\psi(x) = v_t, \quad \psi\left(\frac{\pi}{2} - x\right) = v_{t+1},$$

we have

$$u_{t+1} + u_t = \frac{\pi}{2},$$

$$v_{t+1} + v_t = 1.$$

The solutions of which are

$$u_t = c_1 (-1)^x + \frac{\pi}{4},$$

$$v_t = c_2 (-1)^x + \frac{1}{2}.$$

Hence

$$\frac{v_t - \dfrac{1}{2}}{u_t - \dfrac{\pi}{4}} = \frac{c_2}{c_1} = C.$$

Therefore

$$v_t = \frac{1}{2} + C\left(u_t - \frac{\pi}{4}\right),$$

or

$$\psi(x) = \frac{1}{2} + C\left(x - \frac{\pi}{4}\right).$$

Therefore

$$\phi(x) = \left\{\frac{1}{2} + C\left(x - \frac{\pi}{4}\right)\right\}^{\frac{1}{2}},$$

in which C must be interpreted as a function of x which does not change when x is changed into $\frac{\pi}{2} - x$. It is in fact *an arbitrary symmetrical function of x and $\frac{\pi}{2} - x$.*

The previous solution (12) is included in this.

For, equating the two values of $\phi(x)$ with a view to determine C, we find

$$C = \frac{\chi(x) - \chi\left(\frac{\pi}{2} - x\right)}{x - \frac{\pi}{4}}$$

$$= \frac{\chi(x)}{x - \frac{\pi}{4}} + \frac{\chi\left(\frac{\pi}{2} - x\right)}{\frac{\pi}{2} - x - \frac{\pi}{4}},$$

which is seen to be symmetrical with respect to x and $\frac{\pi}{2} - x$.

8. There are certain equations, and those of no inconsiderable importance, which involve at once two independent variables in such functional connexion that by differentiation and elimination of one or more of the functional terms, the solution will be made ultimately to depend upon that of a differential equation.

Ex. 14. Representing by $P\phi(x)$ the unknown magnitude of the resultant of two forces, each equal to P, acting in one plane and inclined to each other at an angle $2x$, it is shewn by Poisson (*Mécanique*, Tom. I. p. 47) that on certain assumed principles, viz. the principle that the order in which forces are combined into resultants is indifferent—the principle of (so-called) sufficient reason, &c., the following functional equation will exist independently of the particular values of x and y, viz.

$$\phi(x+y) + \phi(x-y) = \phi(x)\,\phi(y).$$

Now, differentiating twice with respect to x, we have

$$\phi''(x+y) + \phi''(x-y) = \phi''(x)\,\phi(y).$$

And differentiating the same equation twice with respect to y,

$$\phi''(x+y) + \phi''(x-y) = \phi(x)\,\phi''(y).$$

Hence

$$\frac{\phi''(x)}{\phi(x)} = \frac{\phi''(y)}{\phi(y)}.$$

Thus the value of $\dfrac{\phi''(x)}{\phi(x)}$ is quite independent of that of x.
We may therefore write

$$\frac{\phi''(x)}{\phi(x)} = \pm\, m^2,$$

m being an arbitrary constant. The solution of this equation is

$$\phi(x) = A\epsilon^{mx} + B\epsilon^{-mx}, \quad \text{or} \quad \phi(x) = A\cos mx + B\sin mx.$$

Substituting in the given equation to determine A and B, we find

$$\phi(x) = \epsilon^{mx} + \epsilon^{-mx}, \quad \text{or} \quad 2\cos mx.$$

Now assuming, on the afore-named principle of sufficient reason, that three equal forces, each of which is inclined to the two others at angles of 120°, produce equilibrium, it follows that $\phi\left(\dfrac{\pi}{3}\right) = 1$. This will be found to require that the second form of $\phi(x)$ be taken, and that m be made equal to 1. Thus $\phi(x) = 2\cos x$. And hence the known law of composition of forces follows.

Ex. 15. A ball is dropped upon a plane with the intention that it shall fall upon a given point, through which two perpendicular axes x and y are drawn. Let $\phi(x)\,dx$ be the probability that the ball will fall at a distance between x and $x + dx$ from the axis y, and $\phi(y)\,dy$ the probability that it will fall at a distance between y and $y + dy$ from the axis x. Assuming that the tendencies to deviate from the respective axes are independent, what must be the form of the function $\phi(x)$ in order that the probability of falling upon any particular point of the plane may be independent of the position of the rectangular axes? (Herschel's *Essays*).

The functional equation is easily found to be

$$\phi(x)\,\phi(y) = \phi\{\sqrt{(x^2 + y^2)}\}\,\phi(0).$$

Differentiating with respect to x and with respect to y, we have

$$\phi'(x)\,\phi(y) = \frac{x\phi'\{\sqrt(x^2+y^2)\}\,\phi(0)}{\sqrt(x^2+y^2)},$$

$$\phi(x)\,\phi'(y) = \frac{y\phi'\{\sqrt(x^2+y^2)\}\,\phi(0)}{\sqrt(x^2+y^2)}.$$

Therefore

$$\frac{\phi'(x)}{x\phi(x)} = \frac{\phi'(y)}{y\phi(y)}.$$

Hence we may write

$$\frac{\phi'(x)}{x\phi(x)} = 2m,$$

a differential equation which gives

$$\phi(x) = C\epsilon^{mx^2}.$$

The condition that $\phi(x)$ must diminish as the absolute value of x increases shews that m must be negative. Thus we have

$$\phi(x) = C\epsilon^{-h^2x^2}.$$

EXERCISES.

1.   If $\phi(x) = \dfrac{2x}{1-x^2}$, determine $\phi^n(x)$.

2.   If $\phi(x) = 2x^2 - 1$, determine $\phi^n(x)$.

3.   If $\psi(t) = \dfrac{a+bt}{c+et}$ and $\psi^x(t) = \dfrac{A+Bt}{C+Et}$, shew, by means of the necessary equation $\psi\psi^x(t) = \psi^x\psi(t)$, that

$$\frac{A}{a} = \frac{E}{e} = \frac{C-B}{c-b}.$$

4. Shew hence that $\psi^x(t)$ may be expressed in the form

$$\frac{a+b_x t}{b_x - b + c + et},$$

the equation for determining b_x being

$$b_x b_{x+1} + cb_{x+1} - bb_x - ae = 0,$$

and that results equivalent to those of Ex. 5, Art. 2, may hence be deduced.

5. Solve the equation $f(x) + f(y) = f(x+y)$.

6. Find the value, to x terms, of the continued fraction

$$\cfrac{2}{1+\cfrac{2}{1+\&c.}}$$

7. What particular solution of the equation

$$f(x) + f\left(\frac{1}{x}\right) = a$$

is deducible by the method of Art. 7 from the equation

$$f(x) + mf\left(\frac{1}{x}\right) = a + n\phi(x).$$

8. Required the equation of that class of curves in which the product of any two ordinates, equidistant from a certain ordinate whose abscissa a is given, is equal to the square of that abscissa.

9. If πx be a periodical function of x of the nth degree, shew that there will exist a particular value of $f(\pi) x$ expressible in the form

$$a_0 + a_1 \pi x + a_2 \pi^2 x \dots + a_{n-1} \pi^{n-1} x,$$

and shew how to determine the constants $a_0, a_1, a_2 \dots a_{n-1}$.

10. Shew hence that a particular integral of the equation

$$\phi\left(\frac{1+x}{1-x}\right) - a\phi(x) = x$$

will be

$$\phi(x) = \frac{a^3}{1-a^4}\left(x + \frac{1}{a}\frac{1+x}{1-x} - \frac{1}{a^2 x} + \frac{1}{a^3}\frac{x-1}{x+1}\right).$$

11. The complete solution of the above equation will be obtained by adding to the particular value of x the complementary function $Ca^{\frac{4\tan^{-1}x}{\pi}}$.

12. Solve the simultaneous functional equations

$$\phi(x+y) = \phi(x) + \frac{\phi(y)\{\psi(x)\}^2}{1 - \phi(x)\phi(y)},$$

$$\psi(x+y) = \frac{\psi(x)\psi(y)}{1 - \phi(x)\phi(y)}.$$

(*Smith's Prize Examination*, 1860.)

CHAPTER XII.

GEOMETRICAL APPLICATIONS.

1. THE determination of a curve from some property connecting points separated by finite intervals usually involves the solution of an equation of differences, pure or mixed, or more generally of a functional equation.

The particular species of this equation will depend upon the law of succession of the points under consideration, and upon the nature of the elements involved in the expression of the given connecting property.

Thus if the abscissæ of the given points increase by a constant difference, and if the connecting property consist merely in some relation between the successive ordinates, the determination of the curve will depend on the integration of a pure equation of differences. But if, the abscissæ still increasing by a constant difference, the connecting property consist in a relation involving such elements as the tangent, the normal, the radius of curvature, &c., the determining equation will be one of mixed differences.

If, instead of the abscissa, some other element of the curve is supposed to increase by a constant difference, it is necessary to assume that element as the independent variable. But when no obvious element of the curve increases by a constant difference, it becomes necessary to assume as independent variable the index of that operation by which we pass from point to point of the curve, i. e. some number which is supposed to measure the frequency of the operation, and which increases by unity as we pass from any point to the succeeding point. Then we must endeavour to form two equations of differences, pure or mixed, one from the law of succession of the points, the other from their connecting property; and from the integrals eliminate the new variable.

There are problems in the expression of which we are led to what may be termed functional differential equations, i. e. equations in which the operation of differentiation and an unknown functional operation seem inseparably involved. In some such cases a procedure similar to that employed in the solution of Clairaut's differential equation enables us to effect the solution.

2. The subject can scarcely be said to be an important one, and a single example in illustration of each of the different kinds of problems, as classified above, may suffice.

Ex. 1. To find a curve such that, if a system of n right lines, originating in a fixed point and terminating in the curve, revolve about that point making always equal angles with each other, their sum shall be invariable. (Herschel's *Examples*, p. 115).

The angles made by these lines with some fixed line may be represented by

$$\theta, \ \theta + \frac{2\pi}{n}, \quad \theta + \frac{4\pi}{n}, \dots \ \theta + \frac{2(n-1)\pi}{n}.$$

Hence, if $r = \phi(\theta)$ be the polar equation of the curve, the given point being pole, we have

$$\phi(\theta) + \phi\left(\theta + \frac{2\pi}{n}\right) \dots + \phi\left\{\theta + \frac{2(n-1)\pi}{n}\right\} = na,$$

a being some given quantity.

Let $\theta = \frac{2\pi z}{n}$, and let $\phi\left(\frac{2\pi z}{n}\right) = u_z$, then we have

$$u_z + u_{z+1} \dots + u_{z+n-1} = na,$$

the complete integral of which is

$$u_z = a + C_1 \cos \frac{2\pi z}{n} + C_2 \cos \frac{4\pi z}{n} \dots + C_{n-1} \cos \frac{(2n-2)\pi z}{n}.$$

Hence we find

$$r = a + C_1 \cos \theta + C_2 \cos 2\theta \ldots + C_{n-1} \cos (n-1)\, \theta,$$

the analytical form of any coefficient C_i being

$$C_i = A + B_1 \cos n\theta + B_2 \cos 2n\theta + \&c.,$$
$$+ E_1 \sin n\theta + E_2 \sin 2n\theta + \&c.,$$

$A, B_1, E_1,$ &c., being absolute constants.

The particular solution $r = a + b \cos \theta$ gives, on passing to rectangular co-ordinates,

$$(x^2 - bx + y^2)^2 = a^2 (x^2 + y^2),$$

and the curve is seen to possess the property that " if a system of any number of radii terminating in the curve and making equal angles with each other be made to revolve round the origin of co-ordinates their sum will be invariable."

Ex. 2. Required the curve in which, the abscissæ increasing by a constant value unity, the subnormals increase in a constant ratio $1 : a$.

Representing by y_x the ordinate corresponding to the abscissa x, we shall have the equation of mixed differences

$$y_x \frac{dy_x}{dx} - ay_{x-1} \frac{dy_{x-1}}{dx} = 0 \ \ldots\ldots\ldots\ldots\ (1).$$

Let $y_x \dfrac{dy_x}{dx} = u_x$, then

$$u_x - au_{x-1} = 0\, ;$$
$$\therefore\ u_x = Ca^x,$$

whence

$$y_x \frac{dy_x}{dx} = Ca^x \ \ldots\ldots\ldots\ldots\ldots\ (2).$$

Hence integrating we find

$$y_x = \sqrt{(C_1 a^x + c)} \ \ldots\ldots\ldots\ldots\ldots\ (3),$$

C_1 being a periodical constant which does not vary when x changes to $x+1$, and c an absolute constant.

Ex. 3. Required a curve such that a ray of light proceeding from a given point in its plane shall after two reflections by the curve return to the given point.

The above problem has been discussed by Biot, whose solution as given by Lacroix (*Diff. and Int. Calc.* Tom. III. p. 588) is substantially as follows:

Assume the given radiant point as origin; let x, y be the co-ordinates of the first point of incidence on the curve, and x', y' those of the second. Also let $\dfrac{dy}{dx} = p$, $\dfrac{dy'}{dx'} = p'$.

It is easily shewn that twice the angle which the normal at any point of the curve makes with the axis of x is equal to the sum of the angles which the incident, and the corresponding reflected ray at that point make with the same axis.

Now the tangent of the angle which the incident ray at the point x, y makes with the axis of x is $\dfrac{y}{x}$. The tangent of the angle which the normal makes with the axis of x is $-\dfrac{1}{p}$, and the tangent of twice that angle is

$$\frac{-\dfrac{2}{p}}{1 - \dfrac{1}{p^2}} = \frac{2p}{1 - p^2}.$$

Hence the tangent of the angle which the ray reflected from x, y makes with the axis of x is

$$\frac{\dfrac{2p}{1 - p^2} - \dfrac{y}{x}}{1 + \dfrac{2p}{1 - p^2}\dfrac{y}{x}} = \frac{2xp - y(1 - p^2)}{x(1 - p^2) + 2py} \quad \ldots\ldots\ldots\ldots (1).$$

Again, by the conditions of the problem a ray incident from the origin upon the point x', y' would be reflected in the *same*

straight line, only in an opposite direction. But the two
expressions for the tangent of inclination of the reflected ray
being equal,

$$\frac{2x'p' - y'(1 - p'^2)}{x'(1 - p'^2) + 2y'p'} - \frac{2xp - y(1 - p^2)}{x(1 - p^2) + 2yp} = 0 \ldots\ldots\ldots (2),$$

while for the equation of that ray, we have

$$y' - y = \frac{2xp - y(1 - p^2)}{x(1 - p^2) + 2yp}(x' - x) \ldots\ldots\ldots\ldots (3).$$

Now, regarding x and y as functions of an independent
variable z which changes to $z + 1$ in passing from the first
point of incidence to the second, the above equations become

$$\Delta \frac{2xp - y(1 - p^2)}{x(1 - p^2) + 2yp} = 0,$$

$$\Delta y = \frac{2xp - y(1 - p^2)}{x(1 - p^2) + 2yp} \Delta x.$$

The first of these equations gives

$$\frac{2xp - y(1 - p^2)}{x(1 - p^2) + 2yp} = C \ldots\ldots\ldots\ldots\ldots (4),$$

whence by substitution

$$\Delta y = C\Delta x.$$

Therefore

$$y = Cx + C'.$$

Here C and C' are primarily periodic functions of z which
do not change when z becomes $z + 1$. Biot observes that, if
C be such a function, $\phi(C)$, in which the form of ϕ is arbitrary,
will also be such, and that we may therefore assume $C' = \phi(C)$,
whence

$$y = Cx + \phi(C),$$

and, restoring to C its value in terms of x, y, and p given in
(4), we shall have

$$y = x \frac{2xp - y\,(1 - p^2)}{x\,(1 - p^2) + 2yp} + \phi \left\{ \frac{2xp - y\,(1 - p^2)}{x\,(1 - p^2) + 2yp} \right\}^{*} \ \ldots\ldots \ (5).$$

This is the differential equation of the curve.

Although Lacroix does not point out any restriction on the form of the function ϕ, it is clear that it cannot be quite arbitrary. For if $C = \psi(z)$, we should have

$$C' = \phi\psi\,(z),$$

and then, giving to ϕ some functional form to which ψ is inverse, there would result

$$C' = z,$$

so that C' would change when z was changed into $z + 1$. From the general form of periodic constants, Chap. II., it is evident that a rational function of such a constant possesses the same character. Thus the differential equation (5) is applicable when ϕ indicates a rational function, and generally when it denotes a functional operation which while periodical itself does not affect the periodical character of its subject.

If we make the arbitrary function 0, we have on reduction

$$(y^2 - x^2)\,p + xy\,(1 - p^2) = 0,$$

the integral of which is

$$x^2 + y^2 = r^2,$$

denoting a circle.

---

* It is only while writing this Chapter that a general interpretation of this equation has occurred to me. Its complete primitive denotes a family of curves defined by the following property, viz. that the caustic into which each of these curves would reflect rays issuing from the origin would be identical with the envelope of the system of straight lines defined by the equation $y = cx + \phi\,(c)$, c being a variable parameter. This interpretation, which is quite irrespective of the form of the function ϕ, confirms the observation in the text as to the necessity of restricting the form of that function in the problem there discussed. I regret that I have not leisure to pursue the inquiry.

I have also ascertained that the differential equation always admits of the following particular solution, viz.

$$(y - A)^2 + (x - B)^2 = 0,$$

A and B being given by the equation

$$\phi\,(\sqrt{-1}) = A - B\,\sqrt{-1}.$$

If we make the arbitrary function a constant and equal to $2a$, we find on reduction

$$\{x^2 - (y - a)^2 + a^2\}\, p - x\, (y - a)\, (1 - p^2) = 0,$$

the complete primitive of which (*Diff. Equations*, p. 135), is

$$(y - a)^2 + c^2 x^2 = \frac{a^2 c^2}{1 - c^2},$$

the equation of an ellipse about the focus.

3. The following once famous problem engaged in succession the attention of Euler, Biot, and Poisson. But the subjoined solution, which alone is characterized by unity and completeness, is due to the late Mr Ellis, *Cambridge Journal*, Vol. III. p. 131. It will be seen that the problem leads to a functional differential equation.

Ex. 4. Determine the class of curves in which the square of any normal exceeds the square of the ordinate erected at its foot by a constant quantity a.

If $y^2 = \psi\, (x)$ be the equation of the curve, the subnormal will be $\dfrac{\psi'\, (x)}{2}$, and the normal squared $\psi\, (x) + \left\{\dfrac{\psi'\, (x)}{2}\right\}^2$. The equation of the problem will therefore be

$$\psi\, (x) + \left\{\frac{\psi'\, (x)}{2}\right\}^2 - \psi\left\{x + \frac{\psi'\, (x)}{2}\right\} = a \;\ldots\ldots\ldots\; (1).$$

Differentiating, we have

$$\psi'\, (x) + \psi'\, (x)\, \frac{\psi''\, (x)}{2} + \psi'\left\{x + \frac{\psi'\, (x)}{2}\right\}\left(1 + \frac{\psi''\, (x)}{2}\right) = 0,$$

which is resolvable into the two equations,

$$1 + \frac{\psi''\, (x)}{2} = 0 \;\ldots\ldots\ldots\ldots\ldots\; (2),$$

$$\psi'\, (x) + \psi'\left\{x + \frac{\psi'\, (x)}{2}\right\} = 0 \;\ldots\ldots\ldots\ldots\; (3).$$

The first of these gives on integration

$$\psi(x) + x^2 = ax + \beta \quad\quad\quad\quad (4).$$

Substituting the value of $\psi(x)$, hence deduced, in (1), we find as an equation of condition

$$a = 0,$$

and, supposing this satisfied, (4) gives

$$y^2 + x^2 = ax + \beta,$$

the equation of a circle whose centre is on the axis of x. It is evident that this is a solution of the problem, supposing

$$a = 0.$$

To solve the second equation (3), assume

$$x + \tfrac{1}{2}\psi'(x) = \chi(x),$$

and there results

$$(\chi)^2 x - 2\chi(x) + x = 0 \quad\quad\quad\quad (5).$$

To integrate this let $x = u_t$, $\chi(x) = u_{t+1}$, and we have

$$u_{t+2} - 2u_{t+1} + u_t = 0,$$

whence

$$u_t = C + C't,$$

C and C' being functions which do not change on changing t into $t+1$. If we represent them by $P(t)$ and $P_1(t)$, we have

$$u_t = P(t) + tP_1(t),$$

$$u_{t+1} = P(t) + (t+1)P_1(t),$$

whence,   since $u_t = x$ and $u_{t+1} = \chi(x) = x + \tfrac{1}{2}\psi'(x)$,

we have

$$x = P(t) + tP_1(t),$$

$$\tfrac{1}{2}\psi'(x) = P_1(t),$$

Hence
$$\psi'(x)\,dx = P_1(t)\{P'(t) + P_1(t) + tP_1'(t)\}dt,$$
$$\psi(x) = \int P_1 t\{P'(t) + P_1(t) + tP_1'(t)\}\,dt.$$

Replacing therefore $\psi(x)$ by y^2, the solution is expressed by the two equations,

$$\left.\begin{array}{l} x = P(t) + tP_1(t) \\ y^2 = \int P_1(t)\{P'(t) + P_1(t) + tP_1'(t)\}\,dt \end{array}\right\} \ \ldots\ldots\ (6),$$

from which, when the forms of $P(t)$ and $P_1(t)$ are assigned, t must be eliminated.

If we make $P(t) = a$, $P_1(t) = \beta$, thus making them constant, we have

$$x = a + \beta t,$$
$$y^2 = \int \beta^2\,dt = \beta^2 t + c.$$

Therefore eliminating t and substituting e for $c - a\beta$,

$$y^2 = \beta x + e.$$

Substituting this in (1), we find

$$\frac{-\beta^2}{4} = a.$$

Thus, in order that the solution should be real, a must be negative. Let $a = -h^2$, then $\beta = \pm 2h$, and

$$y^2 = \pm 2hx + e \ \ldots\ldots\ldots\ldots\ldots\ldots\ldots\ (7),$$

the solution required. This indicates two parabolas.

If $a = 0$, the solution represents two straight lines parallel to the axis of x.

EXERCISES.

1. FIND the general equation of curves in which the diameter through the origin is constant in value.

2. Find the general equation of the curve in which the product of two segments of a straight line drawn through a fixed point in its plane to meet the curve shall be invariable.

3. If in Ex. 4 of the above Chapter the radiant point be supposed infinitely distant, shew that the equation of the reflecting curve will be of the form

$$y = \frac{2px}{1-p^2} + \phi\left(\frac{2p}{1-p^2}\right),$$

ϕ being restricted as in the Example referred to.

4. If a curve be such that a straight line cutting it perpendicularly at one point shall also cut it perpendicularly at another, prove that the differential equation of the curve will be

$$y = \frac{-x}{p} + \phi\left(\frac{-1}{p}\right),$$

ϕ being restricted as in Ex. 4 of this Chapter.

5. Shew that the integral of the above differential equation, when the form of ϕ is unrestricted, may be interpreted by the system of involutes to the curve which is the envelope of the system of straight lines defined by the equation

$$y = mx + \phi(m),$$

m being a variable parameter.

MISCELLANEOUS EXAMPLES.

Selected chiefly from the Senate-House Examination Papers.

1.   PROVE the following properties of differences of 0 :

(1) $f(\Delta)\,0^{n+1} = (n+1)\,\dfrac{f(\Delta)}{\log(1+\Delta)}\,0^n.$

(2) $f(\Delta)\,0^{n+1} = (1+\Delta)\,f'(\Delta)\,0^n.$

2.   If $f(t) = \pm f\!\left(\dfrac{1}{t}\right) + C,$ shew that

$$f(D)\,0^{2x+1} = 0, \text{ or } f(D)\,0^{2x} = 0,$$

according as the upper or lower sign is taken.

3.   Shew that

$$f(D)\,a^x \cos mx = ra^x \cos(mx + \theta)\,;$$

where

$$r\cos\theta = f(D)\,a^0 \cos(0\,.\,m),$$

$$r\sin\theta = f(D)\,a^0 \sin(0\,.\,m),$$

in which D operates on 0.

4.   If $S_r = 1^r + 2^r + \ldots + x^r,$ shew that

$$S_{2n} = \frac{1}{2n+1}\frac{d}{dx}\,S_{2n+1},$$

n being a positive integer.

5.   If $a_0 + a_1 x + a_2 x^2 + \ldots = \phi(x),$ and $u_0,\ u_1,\ u_2,\ \ldots$ be successive values of any function independent of $x,$ shew that

$$a_r u_r x^r + a_{n+r} u_{n+r} x^{n+r} + a_{2n+r} u_{2n+r} x^{2n+r} + \ldots$$

$$= \frac{1}{n}\,[\Sigma\,\{a^{n-r}\phi(ax)\}\,u_0 + \Sigma\,\{a^{n-r+1}\phi'(ax)\}\,\Delta u_0 x + \ldots],$$

when Σ denotes summation with respect to values of a which are roots of the equation $a^n - 1 = 0.$

6. If $u_x = a + hx$, and z_1, z_2, ... z_m be the roots of the equation

$$z^m - p_1 z^{m-1} + p_2 z^{m-2} - \ldots + (-1)^m p_m = 0,$$

prove that

$$u_{x+z_1} u_{x+z_2} \ldots u_{x+z_m} = p_m h^m + (p_{m-1} \Delta . 0 + p_{m-2} \Delta . 0^2 + \ldots) u_x h^{m-1}$$

$$+ (p_{m-2} \Delta^2 . 0^2 + p_{m-3} \Delta^2 . 0^3 + \ldots) \frac{u_x u_{x-1}}{1.2} h^{m-2}$$

$$+ \ldots\ldots\ldots\ldots$$

$$+ (p_{m-r} \Delta^r . 0^r + p_{m-r-1} \Delta^r . 0^{r+1} + \ldots) \frac{u_x u_{x-1} \ldots u_{x-r+1}}{1.2 \ldots r} h^{m-r}$$

$$+ \ldots\ldots\ldots\ldots$$

7. If B_1, B_3, ... be Bernoulli's numbers, prove that

$$1 - \frac{B_1}{1.2} \pi^2 - \frac{B_3}{1.2.3.4} \pi^4 - \ldots = 0.$$

8. Assuming that

$$\frac{z}{\log(1+z)} = 1 + \frac{1}{2} z - \frac{1}{12} z^2 + \frac{1}{24} z^3 - \frac{19}{720} z^4 + \ldots,$$

shew that

$$\int_a^\beta \phi(x)\, dx = \Delta x \left\{ \frac{u_0}{2} + u_1 + u_2 + \ldots + u_{n-1} + \frac{u_n}{2} - \frac{1}{12}(\Delta u_{n-1} - \Delta u_0) \right.$$

$$\left. - \frac{1}{24}(\Delta^2 u_{n-2} + \Delta^2 u_0) - \frac{19}{720}(\Delta^3 u_{n-3} - \Delta^3 u_0) + \ldots \right\},$$

where $\beta - a = n\Delta x$, and u_0, u_1, ... u_n are successive values of $\phi(x)$ at equal intervals from $x = a$ to $x = \beta$.

9. Shew that

$$z_n = \frac{d^n u}{dx^n} \div \frac{d^{n-1} u}{dx^{n-1}}$$

satisfies the equation

$$\frac{dz_n}{dx} = z_n \Delta z_n,$$

u being any function of x.

If a regular polygon, which is inscribed in a fixed circle, be moveable, and if x denote the variable arc between one of its angles and a fixed point in the circumference, and z_n the ratio, multiplied by a certain constant, of the distances from the centre of the feet of perpendiculars drawn from the nth and $(n-1)^{th}$ angles, counting from A, on the diameter through the fixed point, prove that z_n is a function which satisfies the equation.

10.   If $\phi(z) = \phi(x)\,\phi(y)$, where z is a function of x and y determined by the equation $f(z) = f(x)f(y)$, find the form of $\phi(x)$.

ANSWERS TO THE EXERCISES.

II. 11. $\epsilon\left(1 + t + t^2 + \dfrac{5}{6}t^3\right).$

III. 1. 2·3263359, which is true to the last figure.

2. $v_2 = \dfrac{-3v_0 + 10v_1 + 5v_4 - 2v_5}{10}, \quad v_3 = \dfrac{-2v_0 + 5v_1 + 10v_4 - 3v_5}{10}.$

3. $x^2 - x + 4.$

IV. 1. 385.　　　　2. $\dfrac{1}{9} - \dfrac{9x + 16}{6\,(x + 2)\,(x + 3)\,(x + 4)}.$

3. $\dfrac{2}{3} + \dfrac{4^{x+1}}{3} \times \dfrac{x - 1}{x + 2}.$

5. $\dfrac{1 - a\cos\theta + a^{x+1}\cos(x + 1)\,\theta - a^x\cos x\theta}{1 - 2a\cos\theta + a^2}.$

7. Assume for the form of the integral

$$\frac{(A + Bx \ldots + Mx^{n-1})\,s^x}{u_x u_{x+1} \ldots u_{x+m-2}},$$

and then seek to determine the constants.

8. $\cot\dfrac{\theta}{2} - \cot(2^{x-1}\theta).$　　　　11. Refer to p. 62.

12. $\dfrac{\sin m\,(2x-1)}{2\sin m}\,\phi(x) + \dfrac{\cos m\,(2x)}{(2\sin m)^2}\,\Delta\phi(x) - \dfrac{\sin m\,(2x+1)}{(2\sin m)^3}\,\Delta^2\phi(x)$

$- \dfrac{\cos m\,(2x + 2)}{(2\sin m)^4}\,\Delta^3\phi(x) + \dfrac{\sin m\,(2x + 3)}{(2\sin m)^5}\,\Delta^4\phi(x) + \&c.$

V.   1. $\dfrac{1}{a}\dfrac{\pi}{2}$ and $\dfrac{1}{a}\left(\dfrac{\pi}{2}-\tan^{-1}\dfrac{1}{a}\right)$, which, if $a=1$, reduce to $\dfrac{\pi}{2}$ and $\dfrac{\pi}{4}$.

VI.   1. $7\cdot4854709$.   **2.** $2567\cdot6046$.

4. $\pi^{-\frac{1}{2}}\left(2x^{\frac{1}{2}}+\dfrac{3}{4x^{\frac{1}{2}}}\right)$.

5. $\dfrac{1}{2}+\dfrac{1}{2\cdot2^2}+\dfrac{1}{3\cdot2^3}+\dfrac{1}{4\cdot2^4}\ \ldots+$ &c.

VII.   1.   1st. $u=\dfrac{\Delta u}{2x+1}\left(x^2+\dfrac{\Delta u}{2x+1}\right)$.   **2nd.** The same.

3rd. $\left\{\dfrac{1}{(a-1)^2}-\dfrac{x}{a-1}\right\}\Delta^2 u_x+x\Delta u_x-u_x=0$.

4th. $\left(\dfrac{\Delta u}{a-1}\right)^2+a^{2x}\left(\dfrac{\Delta u}{a-1}-u\right)=0$.   **5th.** The same.

2. $u_x=c_1\,(-1)^x+c_2\,(4)^x$.

3. $u_x=c_1\,(-1)^x+c_2\,(4)^x+\dfrac{m^x}{m^2-3m-4}$.

4. $u_x=cp^{x-1}a^{x^2-x}+\dfrac{qa^{(x-1)^2}}{1-ap}$.

5. $u_x=\dfrac{\cos n\,(x-1)-a\cos nx}{a^2-2a\cos n+1}+Ca^x$.

6. $u_x=(c_0+c_1x)\,(-2)^x+\dfrac{x}{9}-\dfrac{2}{27}$.

7.   Note that this may be put in the form

$$u_x-3\sin(\pi x)\,u_{x-1}+2\sin(\pi x)\sin\{\pi\,(x-1)\}\,u_{x-2}=0.$$

The final integral is $u_x=(\sin\pi x)^x\{c_0+c_1\,(2)^x\}$.

9. $u_x = (-1)^x (c_0 + c_1 x) + \dfrac{x(x-1)(x-2)}{4} - \dfrac{3}{4} x(x-1) + \dfrac{9}{8} x - \dfrac{3}{4}$.

10 $u_x = (m^2 + n^2)^{\frac{x}{2}} \left\{ c_1 \cos \left(x \tan^{-1} \dfrac{n}{m} \right) + c_2 \sin \left(x \tan^{-1} \dfrac{n}{m} \right) \right\}$.

11. $u_x = \dfrac{c}{2^x - c} - x$.

13. Express it first in the symbolical form
$$(D - a^{-x})(D - a^x) u = 0.$$

14. $u_x = a^{\frac{x(x-1)}{2}} \{ c + c' \Sigma a^{-\frac{x(x-1)}{2}} \}$.

15. Express it in the form $(D - a)(D - a^{-x}) = 0$.

16. $u_x = \sqrt{a} \tan \left(C_1 \cos \dfrac{2\pi x}{3} + C_2 \sin \dfrac{2\pi x}{3} \right)$.

17. $u_x = A + Ba^x + Ca^{2x}$, where a is an imaginary cube root of unity, and A, B, C are connected by the condition
$$A^3 + B^3 + C^3 = 3A(a + BC).$$

18. Divide by $u_x u_{x+1} u_{x+2}$, and then compare with the last example.

VIII. 1. $y = \dfrac{a+b}{2} x - \dfrac{a-b}{4} (-1)^x + C$. 3. The complete primitive is of the first degree with respect to the constants.

4. $y = C^2 + C \dfrac{1-a}{1+a} (-a)^x - \dfrac{a^{2x+1}}{(1+a)^2}$.

IX. 2. Yes.

X. 1. $u_x = c_1 + c_2 (-1)^x + mx$, $v_x = c_1 - c_2 (-1)^x - m(x+1)$.

2. $u_x = c_0 + c_1 (-1)^x$, $v_x = c_1 - c_0 (-1)^x$.

3. $u_x = A + \left(B \cos \dfrac{\pi x}{3} + C \sin \dfrac{\pi x}{3} \right)(-1)^x + \dfrac{(m+n-2l)x}{3}$,

$$w_x = A - \left\{ B \cos \frac{\pi (x+1)}{3} + C \sin \frac{\pi (x+1)}{3} \right\} (-1)^x + \frac{(l+m-2n)x}{3},$$

$$v_x = A + \left(B \cos \frac{\pi (x+2)}{3} + C \sin \frac{\pi (x+2)}{3} \right) (-1)^x + \frac{(l+n-2m)x}{3}.$$

4.　　$u_{x,y} = a^x \epsilon^{-\frac{y}{a}} \left(\dfrac{d}{dy} \right)^x \phi(y).$

5.　　$u_{x,y} = a^x \left(\dfrac{d}{dy} \right)^x \phi_1(y) + \beta^x \left(\dfrac{d}{dy} \right)^x \phi_2(y),$　where a and β are
roots of $m^2 - am + b = 0.$

6.　　$u_{x,y} = x(y+x-1) + \phi(y+x).$

7.　　$u_{x,y} = \dfrac{a^{x-y}}{a^{1-n}-1} + \phi(y-nx).$

XI.　1.　$\phi^n(x) = \sqrt{-1} \dfrac{(\sqrt{-1}+x)^m - (\sqrt{-1}-x)^m}{(\sqrt{-1}+x)^m + (\sqrt{-1}-x)^m}.$　$(m = 2^n).$

2.　$\phi^n(x) = \dfrac{1}{2} \{ (x+\sqrt{x^2-1})^m + (x-\sqrt{x^2-1})^m \}.$　$(m = 2^n).$

5.　$f(x) = cx.$ 　　　　　　　　　6.　$\dfrac{2^{x+1} - 2(-1)^x}{2^{x+1} + (-1)^x}.$

7.　$f(x) = \dfrac{a}{2} + \phi(x) - \phi\left(\dfrac{1}{x}\right).$

8.　$y = c\epsilon^{\phi(x-a)},$ $\phi(x)$ denoting an even function of x.

9.　Develope $f(\pi)$ in ascending powers of π, and apply the conditions of periodicity.

12.　$\phi(x) = \tan mx.$
　　　　$\psi(x) = \sec mx.$

MISCELLANEOUS EXAMPLES.

7.　For some remarks on this problem see *Quarterly Journal of Mathematics*, Vol. III. p. 262.　8. See De Morgan's *Calculus*, Ch. XIII. Arts. 61, 67.　10.　$\phi(x) = \{f(x)\}^m.$

Printed in the United States
By Bookmasters